A NEW TREATISE

ON

STEAM ENGINEERING

PHYSICAL PROPERTIES OF PERMANENT GASES

AND OF

DIFFERENT KINDS OF VAPOR

BY

JOHN W. NYSTROM, C. E.

NEW YORK

G. P. PUTNAM'S SONS

182 FIFTH AVENUE

1876

PREFACE.

THE object of this treatise is to furnish a variety of matters pertaining to STEAM ENGINEERING which appear to be wanting in that profession, and which have heretofore not been published.

The authors consulted for this work are eminent experimenters, such as Regnault and Rudberg on steam and gases, Faraday, Pelouze and Andrews on carbonic acid, Favre and Silberman on heat of combustion, Kopp on volume of water, Fairbairn and Tate on volume of steam. None of these savans, however, are responsible for the formulas and tables herein deduced from their experiments.

Where physical sciences are not sufficiently developed to establish a law of action mathematically, experiments are made for the purpose of guiding us to the law; but it can rarely ever be expected that experiments alone can give perfect results, but they give an approximation to the law of variation, which must finally be adjusted and established by the aid of mathematics. This is what has been attempted in the present work.

It was at first not intended to include in this work the steam-tables which are published in the author's *Pocket-Book*, but after having carefully investigated the Fairbairn experiments and formula for volume of steam and concluding that they could not be relied upon, it was therefore decided to calculate new steam-tables and extend them to a pressure of 1000 pounds to the square inch.

The relation between temperature and pressure of steam is also slightly altered in the new steam-tables so as to conform to a uniform curve or law, because the average curve adopted by Regnault does not follow a regular law, and therefore indicates that there must have been some inexactness in his experiments.

When the author worked out the first steam-table in the Navy Department under the direction of Chief-engineer Isherwood, the irregularity of the Regnault curve was then demonstrated with attempts to correct it, but the Chief would not allow any deviation from that curve. The difference is, however, within probable experimental errors, and so small that it is not of much importance in practice.

The author believes that the relation between temperature, pressure and volume of steam, as given in these new tables, is nearest right. The old steam-tables are, however, referred to and used in the body of this work for the reason that many readers may have more faith in them than in the new tables, which are equally applicable to the examples.

Many mathematical proofs have been omitted in this work in order to avoid extensive algebraical demonstrations, which are objectionable to the general reader who only needs the resulting formulas for the insertion of his given numerical values.

The principal formulas are accompanied with examples and also tables ranging between practical limits, showing at a glance the relation between and proportion of the operating elements.

The calculus has been resorted to in only a few cases of necessity where the result could not otherwise be reached.

The numbers of the examples are arranged to correspond with the numbers of the formulas, and therefore do not run in order.

Profound and high-sounding terms, like "potential and kinetic energy," etc., are not used in this work, which limits itself to simple terms such as are used in the shop, and which express the true meaning of the respective cases.

The appendix on "Mechanical Terms" is added to this work to furnish an idea of the unsettled condition of that subject.

Similar discussions have been published in pamphlet form and distributed gratis to institutions of learning.

ALPHABETICAL INDEX.

INDEX TO TABLES.

STANDARD NOTATION OF LETTERS.

It has been attempted throughout this work to adopt a standard notation of letters, for which some new characters have been added to distinguish different quantities which have heretofore been denoted by identical letters.

It is of great importance in technical works that the formula should be clear at a glance without special reference to the meaning of its characters.

The characters ⊟, □, 𝑻, 𝒕, ⩔, 𝑽, HP, ℘, Є and ∂ have been made especially for this work.

The letters T and t denote time, 𝑻 and 𝒕 temperature. V and v denote velocity, ⩔ and 𝑽 volume. P and p denote pressure, and ℘ power.

Mr. W. Barnet Le Van proposed the letter ⩔ to denote volume of steam, as a distinction from V, which is used to denote velocity.

Differential is denoted by ∂, and is placed close to its variable quantity, like ∂x (not $\partial\ x$), because the two letters denote only a single quantity.

The common letter d is needed for denoting diameter, distance, depth and other quantities.

The character ∂ is more distinct in denoting the differential, which is not a common notation, and should be conspicuous like the integral $\int \partial x$.

The character ∂ ought not to be used for any other notation but differential.

The special characters ⊟ and □, denoting grate surface and heating surface, are new and explicit for steam-boiler notations.

The characters ℘, denoting weight in pounds per cubic foot, and Є cubic feet per pound, are also explicit notations which ought to be permanently maintained.

STEAM NOTATION.

P = absolute steam-pressure, lbs. per sq. in.

p = steam pressure above that of atmosphere.

𝔳 = steam volume compared with that of its water.

H = units of heat per pound in steam.

H′ = units of heat per cubic foot in steam.

L = latent heat per pound in steam.

L′ = latent heat per cubic foot in steam.

𝔚 = pounds per cubic foot.

= cubic feet per pound.

𝔗 = temperature Fahr. of steam.

J = thermodynamic equivalent.

X = grade of expansion of steam.

WATER NOTATION.

𝔙 = volume of water, that at 39° or 40° = 1.

𝔱 = temperature Fahr. of water.

l = latent heat per pound in water from 32°.

l′ = latent heat per cubic foot of water.

h = units of heat per pound of water.

h′ = units of heat per cubic foot of water.

𝔚 = weight in pounds per cubic foot of water.

𝔈 = fraction of a cubic foot per pound of water.

W = cubic feet of water.

w = cubic inches of water.

lbs. = pounds of water.

DYNAMICAL NOTATIONS.

F = force in pounds avoirdupois.

V = velocity in feet per second.

T = time of action in seconds.

S = *V T*, space in feet or cubic feet.

𝔓 = *F V*, power in effects or second foot-pounds.

HP = 550 𝔓, horse-power, Watt's unit.

K = *F V T*, work in foot-pounds.

STEAM-BOILER NOTATION.

⊟ = area of firegrate in square feet.

□ = area of heating surface in square feet.

D = diameter of boiler in inches.

d = diameter of staybolts in inches.

t = thickness of boiler-plates in inches.

S = breaking-strain per square inch of iron.

H = height of chimney in feet.

A = cross-area of chimney in square feet.

PERMANENT GASES NOTATION.

𝔳 and 𝔙 = volumes.

𝔗 and 𝔱 = actual temperatures.

𝕿 and 𝖙 = ideal temperatures.

P and *p* = absolute pressures.

𝔚 = pound per cubic foot.

h = units of heat.

S = specific heat, constant volume.

s = specific heat, any volume and pressure.

W = weight of gas in pounds.

MECHANICS.

DEFINITIONS OF THE PRINCIPAL TERMS IN MECHANICS.

MECHANICS is that branch of natural philosophy which treats of the three simple physical elements **force, velocity** and **time,** with their combinations, constituting the functions **power, space** and **work.**

Mechanics is divided into two distinct parts—namely, **Statics** and **Dynamics.**

STATICS is the science of forces in equilibrium or at rest.

DYNAMICS is the science of forces in motion, producing power and work.

QUANTITY is any principle or magnitude which can be increased or diminished by augmentation or abatement of homogeneous parts, and which can be expressed by a number.

ELEMENT is an essential principle which cannot be resolved into two or more different principles.

FUNCTION is any compound result or product of two or more different elements.
A function is resolved by dividing it with one or more of its elements.

Force, velocity and **time** are simple physical elements.

Power, space and **work** are functions of those elements.

These six terms represent the principal elements and functions in Mechanics. All creation, work or action, of whatever kind, whether mechanical, chemical or derived from light, heat, electricity or magnetism—all that has been and is to be done or undone—is comprehended by the product of *force, velocity* and *time.*

FORCE is any action which can be expressed simply by weight, without regard to motion, time, power or work. It is an essential principle which cannot be resolved into two or more different principles, and is therefore a simple element.

VELOCITY is speed or rate of motion. It is an essential principle which cannot be resolved into two or more principles, and is therefore a simple element.

TIME is duration or that measured by a clock. It is an essential principle which cannot be resolved into two or more different principles, and is therefore a simple element.

POWER is the product of the first and second elements, **force** and **velocity**, and is therefore a function.

SPACE is the product of the second and third elements, **velocity** and **time**, and is therefore a function.

WORK is the product of the three simple elements **force, velocity** and **time**, and is therefore a function.

Work is also the product of the element *force* and function *space*, because the function space contains the elements *velocity* and *time.*

Work is also the product of the function *power* and element *time*, because the function *power* contains the elements *force* and *velocity.*

MOMENTUMS are of two kinds—namely, **Static** and **Dynamic.**

STATIC-MOMENTUM is the product of **force** and the **lever** upon which it acts, and is therefore a function.

DYNAMIC-MOMENTUM is the product of **mass** and its **velocity**, which is equal to the product of the **force** and **time** that has produced the velocity of the mass, and is therefore a function.

MASS is the real quantity of matter in a body, and is proportionate to weight when compared in one and the same locality. **Mass** is an essential principle which cannot be resolved into two or more principles, and is therefore a simple element.

The new treatise on "**Elements of Mechanics,**" published by Porter & Coates, Philadelphia, gives complete explanations, with practical examples of the mechanical elements and functions.

STATICS.

ALGEBRAICAL AND GEOMETRICAL EXPRESSIONS OF THE FUNDAMENTAL PRINCIPLES OF STATICS.

Levers of Different Kinds.

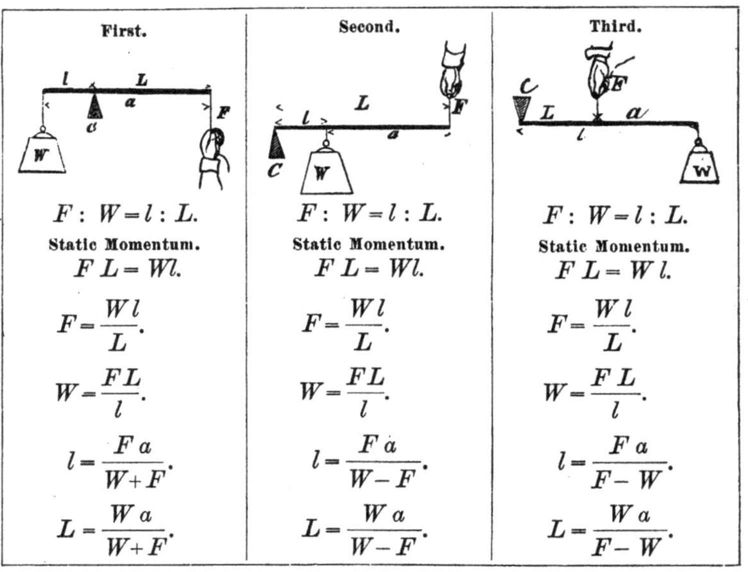

First.	Second.	Third.
$F : W = l : L.$	$F : W = l : L.$	$F : W = l : L.$
Static Momentum.	**Static Momentum.**	**Static Momentum.**
$F\,L = Wl.$	$F\,L = Wl.$	$F\,L = W\,l.$
$F = \frac{W\,l}{L}.$	$F = \frac{W\,l}{L}.$	$F = \frac{W\,l}{L}.$
$W = \frac{F\,L}{l}.$	$W = \frac{F\,L}{l}.$	$W = \frac{F\,L}{l}.$
$l = \frac{F\,a}{W+F}.$	$l = \frac{F\,a}{W-F}.$	$l = \frac{F\,a}{F-W}.$
$L = \frac{W\,a}{W+F}.$	$L = \frac{W\,a}{W-F}.$	$L = \frac{W\,a}{F-W}.$

DYNAMICS.

ALGEBRAICAL AND GEOMETRICAL EXPRESSIONS OF THE FUNDAMENTAL PRINCIPLES OF DYNAMICS.

Elements.	Functions.
Force $= F.$	Power $\mathfrak{P} = F\,V.$
Velocity $= V.$	Space $S = V\,T.$
Time $= T.$	Work $K = F\,V\,T.$
Mass $= M.$	Work $K = \frac{1}{2}\,M\,V^2.$

$F : M = V : T.$

Momentum.

$F\,T = M\,V.$

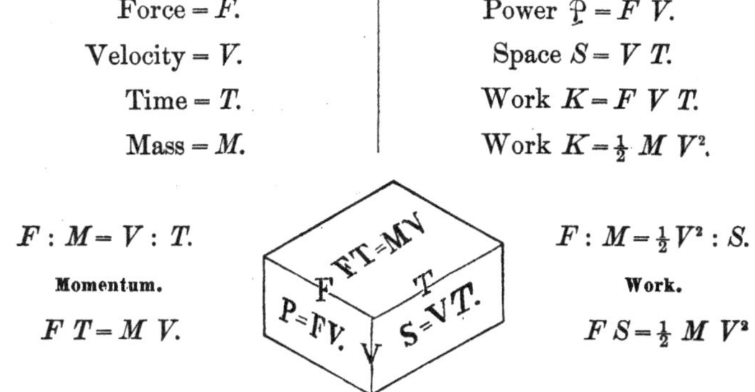

$F : M = \frac{1}{2} V^2 : S.$

Work.

$F\,S = \frac{1}{2}\,M\,V^2$

These are the fundamental principles in Mechanics.

REJECTED TERMS IN MECHANICS.

The author has rejected a great number of terms in Mechanics which are considered useless, confusing and without definite meanings, a list of which is given below and on the next page.

High-sounding terms without definite meaning render the subject of Mechanics difficult to learn, for which reason the author has decided to employ only such terms as are used in the shop.

The language of Mechanics used in schools and text-books differs so much from that used in practice that when a graduate student converses with a practical man on that subject, they do not understand each other, and the latter derides the former as theoretical. This is the principal reason why theoretical sciences are so little available in practice.

In the Appendix to this book is given an example of the language of Mechanics as used in institutions of learning, from which it will be perceived that the author has good reasons for having undertaken a revision of the subject.

The list of rejected terms on the next page is taken from the new treatise of "Elements of Mechanics," to which the following list of expressions and terms is added :

Mechanics of a material point . . .	W. p. 165.
Forces in space	W. p. 182.
Principles of virtual velocity . . .	W. p. 185.
Couples	W. p. 200.
Dynamical stability	W. p. 269.
Modulus of a machine	M.
Intensity of force	W. p. 164.
Strength of impact	W. p. 102.
Intensity of the effort	B. p. 49.
Effort of mechanical work	B. p. 57.
Living force impressed	B. p. 82.
Equilibrium in a knot	W. p. 281.

These kinds of terms and expressions convey no definite meaning, and are not used in practice.

DYNAMICAL TERMS.

Rejected Terms.	Reason for Rejection.
Effort of force.	Means simply force.
Efficiency of force.	" " "
Acting force.	All forces act.
Force of motion.	Means motive force.
Working force.	" " "
Quantity of moving force.	" " "
Quantity of motion.	Has no definite meaning.
Mode of motion.	" " "
Mode of force.	" " "
Moment of activity.	Means simply power.
Mechanical power.	" " "
Mechanical effect.	" " "
Quantity of action.	" " "
Efficiency.	" " "
Rate of work.	" " "
Dynamic effect.	Used for power or work.
Quantity of work.	Means simply work.
Actual total quantity of work.	" " "
Total amount of work.	" " "
Actuated work.	" " "
Vis-viva.	" " "
Living force.	" " "
Energy.	" " "
Actual energy.	" " "
Potential energy.	" " "
Kinetic energy.	" " "
Energy of motion.	" " "
Energy of force.	" " "
Heat a form of energy.	" " "
Heat a mode of motion.	" " "
Mechanical potential energy.	" " "
Quantity of energy.	" " "
Stored energy.	" " "
Intrinsic energy.	" " "
Total actual energy.	" " "
Work of energy.	" " "
Equation of energy.	Formula for work.
Equality of energy.	Primitive and realized work.

STEAM ENGINEERING.

§ 1. A STEAM-ENGINE is only a tool by which the power generated in the steam-boiler is transmitted to where the work is executed, like a water-wheel which transmits the power of a waterfall to its destination.

In hydraulics we define correctly the power of a waterfall, which is called "*the natural effect* of the fall," in distinction from the power transmitted by the water-wheel; but in steam engineering we have heretofore not defined correctly the natural effect generated in the steam-boiler as distinct from that transmitted by the engine.

A badly-constructed water-wheel may transmit only twenty per cent. of the natural effect of the waterfall, whilst a properly-constructed wheel may transmit as high as eighty per cent. or more of the power of the fall. Such is the case also with steam-engines. A badly-constructed steam-engine transmits a much smaller percentage of the natural effect from the boiler than does a better constructed engine. Therefore the power obtained by indicator diagrams from the engine is not a correct measure of the power or steaming capacity of the boiler.

§ 2. From experimental data we have given the volume of steam generated by the evaporation of a given volume of water, which steam volume multiplied by the steam pressure, gives the work done by the steam. This work divided by the time in which it is executed, gives the natural effect or power of the evaporation, independent of the power transmitted by the steam-engine, supposing that the steam is fully admitted throughout the stroke of the piston.

When the steam is expanded in the steam-cylinder, the above defined power multiplied by 1 + the hyperbolic logarithm for the expansion, gives the natural effect of the steam.

§ 3. The primary source of power is derived from the combustion of fuel in the furnace generating heat which penetrates the heating surface into the water which is thus evaporated.

The act of combustion is power, which, multiplied by time, is work.

The act of evaporation is power, which, multiplied by time, is work.

The natural effect or power of combustion is not wholly transmitted to evaporation, but part of it escapes through the chimney.

The physical constitution of heat is not yet well understood, for which reason we cannot give an intelligent explanation of the dynamic elements of combustion and evaporation; but one thing appears to be certain—namely, that the *temperature* of the heat represents *force*, which is the origin of all power and work. It is also known and demonstrated that heat is convertible into work; and consequently, heat must be the product of the three simple physical elements, *force*, *velocity* and *time*.

If the *temperature* of the heat represents *force*, then the space occupied by the heat must evidently represent the product of *velocity* and *time*.

Here it is necessary to refer the reader to the author's *New Treatise on Elements of Mechanics*, published by Porter & Coates, Philadelphia.

Fig. 1.

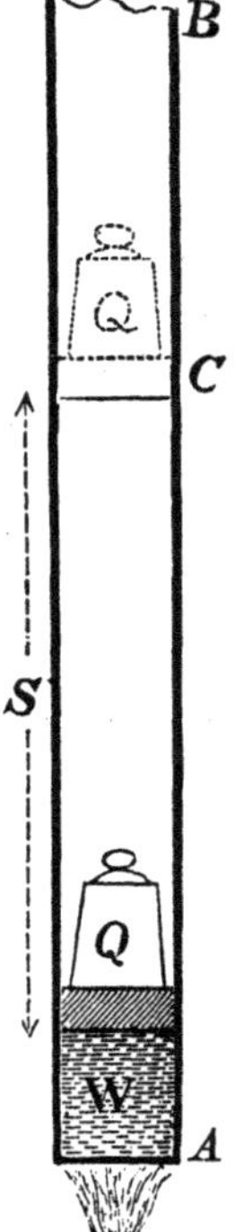

§ 4. The expression "*horse-power of a steam-boiler*" is understood to mean the horse-power of evaporation in the boiler, which power is derived from the heat in the furnace.

For simplicity of illustration, let the steam-boiler be represented by the tube *A B*, of one square foot section, with a bottom at *A* and open at the top *B*.

One cubic foot of water *W* is placed on the bottom in the tube and covered with a tight piston loaded with a weight *Q*.

A burning lamp *L* is placed under the bottom to heat the water for making steam.

The steam-pressure thus generated will raise the piston with the weight *Q* to a height *S*, and the work accomplished by the steam will be the weight *Q* (which must include the pressure of the atmosphere on one square foot, and also the weight of the piston, which is supposed to move without friction) multiplied by the height *S* which the piston is raised. This work divided by the time in which it is accomplished, gives the power of evaporation, which is generally termed the power of the boiler.

Assume the steam-pressure to be 100 pounds to the square inch above vacuum, then $100 \times 144 = 14400$ pounds, the required weight of *Q*. When all the water —that is, one cubic foot—is evaporated, the steam volume will be 267.8 cubic feet; and as the section of the tube is one square foot, the piston must have been lifted 267.8 feet, minus the one foot occupied by the water, or $S = 266.8$ feet.

The work accomplished by the steam will then be $266.8 \times 14400 = 3{,}831{,}920$ foot-pounds.

Suppose this work to be accomplished in the time of one minute, and the power of the evaporation will be,

$$\frac{3831920}{33000} = 116.12 \text{ horse-power.}$$

This should be the natural effect of the steam without expansion.

§ 5. Now, diminish the weight Q gradually, so as to allow the steam to expand—say to double its volume. Then, the hyperbolic logarithm for $2 = 0.69315$, multiplied by the primitive horse-power 116.12, gives 80.488 horse-power gained by the expansion alone, and the gross effect of the steam will be $116.12 + 80.488 = 196.608$ horse-power.

It will be noticed that the one cubic foot of steam which displaced the water was lost in the natural effect of the evaporation; and that is the steam-volume required for pumping the feed-water into the boiler in order to maintain a constant height of water-level.

By the aid of algebra the above argument can be made general for any steam-pressure and dimension of boiler, for which we will adopt the following notation of letters:

W = cubic feet of water of temperature 32° Fahr. evaporated in the time T seconds.

P = steam-pressure in pounds per square inch above vacuum.

$\mathfrak{V}$ = volume of steam compared with that of its water at 32° Fahr. This volume can be found in Nystrom's *Pocket-Book*, pages 398, 399, calculated from the formula of Fairbairn and Tate, which is yet the highest authority on that subject.

$\mathfrak{P}$ = power in effects, or second-foot-pounds.

HP = horse-power of evaporation.

S = space generated by the steam in cubic feet.

F = force in pounds.

V = velocity in feet per second.

T = time of operation in seconds.

K = work in foot-pounds done in the time T by the steam.

X = grade of expansion of the steam.

The Fairbairn's formula for the volume of steam compared with water at 32° Fahr. is

$$\mathfrak{V} = 25.62 + \frac{24307}{P + 0.358}.$$

See arguments on dryness and humidity of steam, in regard to Fairbairn's steam-volume.

The space S, generated by the steam in cubic feet, will be

$$S = W(\psi - 1) \quad . \quad . \quad . \quad . \qquad 1$$

§ 6. This space multiplied by the steam-pressure will be the work done by the steam; and as the space or steam-volume is expressed in cubic feet, the steam-pressure must be expressed per square foot, or $144\,P$.

The unit 1 in the factor $(\psi - 1)$ represents the primitive volume occupied by the water evaporated, and which unit of volume is consumed in feeding the boiler with water, as before explained.

The work accomplished by the steam will then be in foot-pounds.

$$K = W(\psi - 1)\,144\,P \quad . \quad . \quad . \qquad 2$$

Work is the product of the three simple physical elements, force F, velocity V and time T, or

$$K = F\,V\,T \quad . \quad . \quad . \quad . \qquad 3$$

Power $\mathfrak{P}$ is the product of the two elements force F and velocity V, or

$$\mathfrak{P} = F\,V \quad . \quad . \quad . \quad . \qquad 4$$

This power is expressed in effects, each of a force of one pound, moving with a velocity of one foot per second, of which there are 550 effects per horse-power, or

$$\mathrm{HP} = \frac{F\,V}{550} \quad . \quad . \quad . \quad . \quad . \qquad 5$$

The formulas 2 and 3 give the work

$$K = W(\psi - 1)\,144\,P = F\,V\,T = \mathfrak{P}\,T \quad . \quad . \qquad 6$$

Work is the product of power and time, and consequently, if we eliminate the time from the work, we obtain the power, or

$$\mathfrak{P} = \frac{K}{T} = \frac{W}{T}(\psi - 1)\,144\,P, \quad . \quad . \quad . \qquad 7$$

of which the horse-power will be

$$\mathrm{HP} = \frac{W}{550\,T}(\psi - 1)\,144\,P \quad . \quad . \quad . \quad . \qquad 8$$

This formula reduces itself to

$$\mathrm{HP} = \frac{W\,P(\psi - 1)}{3.819\,T} \quad . \quad . \quad . \qquad 9$$

This is the natural effect or gross horse-power of evaporation of water into steam without expansion.

§ 7. The quantity of water which must be evaporated under a pressure P in the time T in order to generate a given horse-power will be

$$W=\frac{3.819\,T\,\text{HP}}{P(\mathfrak{V}-1)} \quad . \quad . \quad . \quad . \qquad 10$$

Assuming the quantity of water evaporated per hour as a measure of gross horse-power of evaporation, we have the time $T=3600$ seconds. Then $3.819\times3600=13748.4$. Insert this value for $3.819\ T$ in formula 9, and the gross horse-power of evaporation per hour will be

$$\text{HP}=\frac{W\,P(\mathfrak{V}\ \ 1)}{13748.4} \quad . \quad . \quad . \quad . \qquad 11$$

The quantity of water evaporated per hour per gross horse-power will be

$$W=\frac{13748.4\ \text{HP}}{P(\mathfrak{V}-1)} \quad . \quad . \quad . \quad . \qquad 12$$

Logarithm for $13748.4=4.1382522$.

§ 8. The steam volume $\mathfrak{V}$ is compared with that of water at 32° Fahr.; therefore, in determining the gross horse-power of evaporation of water of a higher temperature, the action must be reduced to that from water at 32°. This reduction is accomplished by the following formula, in which letters denote:

$\mathfrak{t}$ = actual temperature of the feed-water supposed to be higher than 32°.

$\mathfrak{T}$ = temperature of the steam of pressure P.

W = cubic feet of water that would have been evaporated from the temperature 32°.

W' = cubic feet of feed-water evaporated from temperature $\mathfrak{t}$.

$\mathfrak{v}$ = volume of water at temperature t, compared with that at 39°.

$$W=\frac{W'}{\mathfrak{v}}\left(\frac{1082+0.305\ \mathfrak{T}}{1050+\mathfrak{t}+0.305\ \mathfrak{T}}\right) \quad . \quad . \quad . \qquad 13$$

This formula is derived from the units of heat required to evaporate water of temperature 32° to steam of temperature $\mathfrak{T}$.

This reduction is required for comparing the relative steaming capacity of different boilers fed with water of different temperatures. The reduction varies very little for different pressures—namely, from 20 to 150 pounds the difference will show only on the third decimal; for which reason we may practically omit the steam-pressure and calculate the reduction only for different temperatures of the feed-water, as is done in the following Table I.

When the exact relation between pressure, temperature and volume of steam is known, the reduction will likely be independent of the pressure or temperature of the steam. See Humidity of Steam.

TABLE I.

Reduction for Temperature of Feed-water.

Temp. t.	Reduction R.	Logarithm.	Temp. t.	Reduction R.	Logarithm.
40	0.9932	9.9970367	130	0.9105	9.9592620
50	0.9851	9.9934803	140	0.9000	9.9546693
60	0.9761	9.9895039	150	0.8912	9.9499637
70	0.9671	9.9854546	160	0.8815	9.9451979
80	0.9577	9.9812455	170	0.8719	9.9404765
90	0.9486	9.9770612	180	0.8625	9.9357359
100	0.9392	9.9727643	190	0.8529	9.9308916
110	0.9296	9.9683116	200	0.8432	9.9259440
120	0.9199	9.9637468	212	0.8317	9.9199515

§ 9. The actual quantity of feed-water of temperature t, multiplied by the reduction in the table, gives the quantity of water that would have been evaporated when heated from temperature 32° Fahr.

Example 11. A steam-boiler evaporating $W=125$ cubic feet of water per hour under a pressure of $P=75$ pounds to the square inch above vacuum, or 60 pounds above the atmosphere, the temperature of the feed-water being $t=110°$. Required the natural effect or horse-power of the evaporation?

Formula 11. $$\text{HP} = \frac{125 \times 75\,(348.15-1)}{13748.4} = 236.73 \text{ horses.}$$

That is, 0.528 cubic feet of water evaporated per hour per horse-power, or 1.893 horse-power per cubic foot of water evaporated per hour.

Making correction for the temperature of the feed-water 110° (see Table), the horse-power will be $168.53 \times 0.9392 = 220.06$ horse-power, the natural effect of the evaporation.

Example 12. What quantity of water of temperature $t=90°$ must be evaporated under a pressure of $P=90$ pounds to the square inch in order to generate a natural effect of $\text{HP}=150$ horse-power?

Formula 12. $$W = \frac{13748.4 \times 150}{90\,(294.61-1)} = 78.043 \text{ cubic feet.}$$

This volume corrected for temperature gives $78.043 : 0.9486 = 82.275$ cubic feet, the quantity of water required.

TABLE II.

Natural effect of evaporation of water by heat converted into horsepower.

Steam pressure ab. vacm.	Water evaporated per hour per horsepower.			Horse-power per cub. ft.	Equiva-lent work per unit of heat.
	Cubic feet.	Cubic in.	Pounds.		
P	*W*	*w*	*lbs.*	HP	*J*
5	0.6024	1041.0	29.852	1.6600	46.584
10	0.5796	1002.0	28.723	1.7253	48.032
14.7	0.5701	985.2	28.252	1.7540	48.583
20	0.5641	974.7	27.954	1.7727	48.902
25	0.5593	966.5	27.717	1.7879	49.040
30	0.5553	959.6	27.518	1.8008	49.403
35	0.5516	953.2	27.337	1.8130	49.665
40	0.5483	947.4	27.170	1.8238	49.832
45	0.5451	941.9	27.012	1.8345	50.150
50	0.5420	936.6	26.861	1.8450	50.244
55	0.5391	931.5	26.715	1.8549	50.440
60	0.5362	926.6	26.573	1.8649	50.651
65	0.5334	921.6	26.429	1.8747	50.861
70	0.5305	917.1	26.300	1.8850	51.060
75	0.5280	912.5	26.168	1.8936	51.265
80	0.5254	907.9	26.038	1.9033	51.470
85	0.5228	903.5	25.910	1.9127	51.670
90	0.5203	899.1	25.783	1.9219	51.865
95	0.5178	894.7	25.660	1.9312	52.077
100	0.5153	890.5	25.537	1.9406	52.264
105	0.5129	886.2	25.415	1.9497	52.513
110	0.5104	882.0	25.295	1.9592	52.722
115	0.5081	877.9	25.177	1.9681	53.053
120	0.5057	873.8	25.060	1.9774	53.137
125	0.5034	869.8	24.945	1.9865	53.351
130	0.5008	865.3	24.815	1.9968	53.572
135	0.4988	861.9	24.718	2.0048	53.788
140	0.4965	858.0	24.606	2.0140	54.000
145	0.4943	854.1	24.494	2.0230	54.206
150	0.4921	850.4	24.387	2.0321	54.427

The preceding Table II. gives the horse-power per evaporation per hour of water, expressed either in cubic feet, cubic inches or pounds; also the thermo-dynamic equivalent of heat as realized by the steam without expansion.

When the water evaporated is expressed in pounds, the formulas 11 and 12 will appear as follows:

lbs = pounds of water evaporated in the boiler per hour.

$$HP = \frac{\text{lbs}\ P(\not{V} - 1)}{857721} \quad . \quad . \quad . \quad . \quad 14$$

Logarithm for 857721 = 5.9333463.

$$\text{lb} = \frac{857721\ HP}{P(\not{V} - 1)} \quad . \quad . \quad . \quad . \quad 15$$

The correction for temperature of feed-water will be the same by Table I. as when the water is expressed in cubic feet. One cubic foot of water at 32 weighs 62.387 pounds.

Fig. 2.

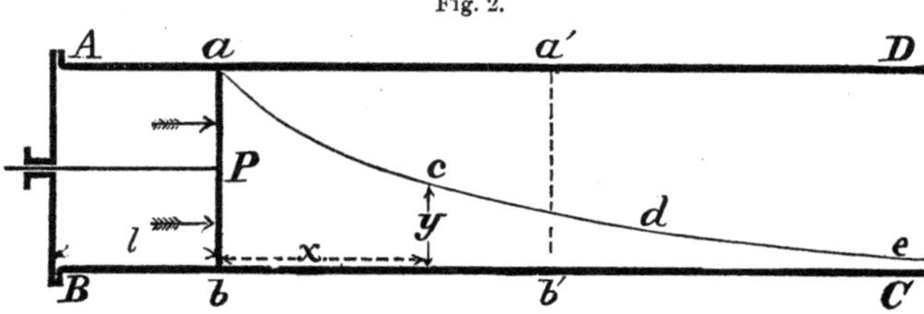

EXPANSION OF STEAM.

§ 10. When steam is working expansively, more power is realized per water evaporated than that given by the Formula 11.

Let *A B C D*, fig. 2, represent a section of a steam cylinder of indefinite length, in which is fitted a piston *a b*, upon which the full steam-pressure *P* is acting in the distance *l*, enclosing the steam-volume *A B a b*, to be expanded. The work accomplished by the full steam-pressure *P* can be represented by the area *A B a b*, or *P l*. When the admittance of steam is cut off, the piston is moved by the expansion of the steam, and the pressure decreases as the steam-volume increases; so that when the volume is doubled the pressure will be one-half or 0.5 *P*, and when the piston has moved two volumes by the expansion—that is, three volumes in all—the pressure will be $\frac{1}{3}P$ at *a′ b′*.

Let the line *A B* represent the axis of ordinates and *B C* the axis of abscissa.

x = distance generated by expansion.
y = ordinate pressure of the expanded steam.

$$\text{Then } P : y = l + x : l \quad . \quad . \quad . \quad 1$$

$$y = \frac{Pl}{l+x} \quad . \quad . \quad . \quad . \quad . \quad 2$$

§ 11. Calculate the ordinate pressure y for several positions of the piston, and set them off as shown in the figure. Join these ordinates by the curve $a\ c\ d\ e$, and the work done by the expansion is represented by the area bounded within that curve and $P\ x\ y$.

k = area, or work of expansion alone, expressed in units of $P\ l$, the work done by the full steam-pressure.

Then $$\partial k = y\ \partial x = \frac{P\ l\ \partial x}{l+x} \quad . \quad . \quad . \quad . \quad 3$$

We have assumed $P\ l$ as unit for the measurement, in which case $P=1$ and $l=1$, and the differential work will be

$$\partial k = \frac{\partial x}{1+x} \quad . \quad . \quad . \quad . \quad . \quad 4$$

$$k = \int \frac{\partial x}{1+x} = hyp.log.\ (1+x) \quad . \quad 5$$

The factor $(1+x)$ represents the whole motion of the piston, of which x is the portion worked with expansion.

s = whole stroke of the piston.
l = part of the stroke worked with full steam.
X = grade of expansion—that is, when the steam is expanded to double its volume, then $X=2$; when three times the volume, $X=3$, and so on.

$$X = \frac{s}{l} = (1+x) \quad . \quad . \quad . \quad . \quad 6$$

The work done by the expansion will then be

$$k = hyp.log.X = hyp.log.\frac{s}{l} \quad . \quad . \quad . \quad 7$$

That is to say, the effect gained by the expansion is equal to the hyperbolic logarithm for the expansion.

When the steam is expanded say four times, then hyp. log. $4 = 1.38629$, or the gain will be 138 per cent. over the effect of that worked with full steam, and the gross effect K will be 238 per cent.

$$K = 1 + hyp.log.X = 1 + hyp.log.\frac{s}{l} \quad . \quad . \quad 8$$

The natural effect or horse-power of evaporation without expansion is

$$HP = \frac{W \cdot P(\psi - 1)}{13748.4} \quad . \quad . \quad . \quad . \qquad 9$$

which multiplied by $(1 + hyp.log.\ X)$, will be the natural effect or horse-power of evaporation with expansion, or

$$HP = \frac{W P (\psi - 1)(1 + hyp.log.X)}{1348.4} \quad . \quad . \qquad 10$$

§ 12. This formula gives the natural effect of evaporation of water into steam, and which, divided into the power given out or transmitted by a steam-engine, gives the efficiency of that steam-engine, as the natural effect of a waterfall divided into the power transmitted by the wheel gives the efficiency of that water-wheel. A compound engine working with a high degree of expansion and condensation of the steam may utilize or transmit as high as 80 per cent. of the natural effect of the steam, whilst a high-pressure or non-condensing engine working against atmospheric pressure may transmit only 40 per cent. of the natural effect.

The expansion X in compound engines is equal to the volume of full steam in the small cylinder, divided into the cubic content of both cylinders.

The cubic content of one steam-port in the small cylinder should be included in the volume of full steam, and the cubic content of one steam-port of each cylinder should be included in the volume of the two cylinders.

Example 10. A set of steam-boilers, evaporating $W = 640$ cubic feet of water per hour, under a pressure of $P = 65$ pounds to the square inch, supply steam to a compound engine in which the steam is expanded $X = 8$ times. Required the natural effect of the steam?

$Hyp.log.8 = 2.07944$. $\psi = 397.51$.

$$HP = \frac{640 \times 65 \times 396.51 \times 3.07944}{13748.4} = 3694.6$$ horse-power, the natural effect required.

It is supposed in this example that the temperature of the feed-water was 32°, for which there is no reduction. The water evaporated per hour per horse-power, in this example, is 0.1723 cubic feet, or 5.773 horse-power per cubic feet evaporated per hour.

EFFECT OF ATMOSPHERIC PRESSURE OPPOSING THE NATURAL EFFECT OF THE STEAM.

§ 13. The volume of air displaced by the steam will be

$$W(\Psi - 1)\frac{s}{l} = W(\Psi - 1)X \quad . \quad . \quad . \quad 1$$

This volume, multiplied by the atmospheric pressure per square foot, will be the work of resistance of the atmosphere, or

$$W(\Psi - 1)X \times 14.7 \times 144 \quad . \quad . \quad . \quad 2$$

That is, 2116.8 $WX(\Psi - 1)$ per hour.

This work, divided by 550×3600 seconds, gives the horse-power of its execution, or

$$\text{HP} = \frac{2116.8\ WX(\Psi - 1)}{550 \times 3600} = \frac{WX(\Psi - 1)}{935.37} \quad . \quad . \quad 3$$

This horse-power, subtracted from Formula 10, will give the natural effect of the steam above that of the atmosphere, or

$$\text{HP} = \frac{WP(\Psi - 1)\ (1 + hyp.log.X)}{13748.4} - \frac{WX(\Psi - 1)}{935.37} \quad . \quad 4$$

$$\text{HP} = \frac{W(\Psi - 1)}{935.37}\left(\frac{P(1 + hyp.log.X)}{14.698} - X\right) \quad . \quad 5$$

This should be the natural effect of steam working through a non-condensing engine, which, divided into the indicated horse-power, gives the efficiency of the motor.

Example 5. A steam-boiler evaporating $W = 85$ cubic feet of water per hour, under a pressure of $P = 100$ pounds to the square inch, supplies steam to a non-condensing engine, cutting off at one-third the stroke, making $X = 3$ the expansion, the temperature of the feed-water being $t = 120°$ Fahr. Required the natural effect of the steam above that of the atmosphere?

*Hyp.log.*3 = 1.0986. $\Psi = 267.8$.

$$\text{HP} = \frac{85 \times 266.8}{935.37}\left(\frac{100 \times 2.0986}{14.698} - 3\right) = 273.37.$$

Correction for temperature of feed-water $t = 120°$. $273.37 \times 0.9199 = 251.48$ horse-power—that is, 0.338 cubic feet of water evaporated per hour per horse-power, or 2.958 horse-power per cubic foot of water evaporated per hour.

MEAN PRESSURE.

§ 14. When the steam is expanded in the cylinder, the mean pressure throughout the stroke of piston will be less than the initial pressure.

F = mean pressure in pounds per square inch.

P = initial pressure.

X = grade of expansion.

s = length of stroke in inches.

l = part of stroke with full steam, in inches.

The mean pressure during the expansion will be $\frac{Pl}{s}$ *hyp.log.X*, which, added to $\frac{Pl}{s}$, gives the mean pressure throughout the stroke, or

$$F = \frac{Pl}{s} + \frac{Pl}{s}\, hyp.log.X \quad . \quad . \quad . \quad 1$$

$X = \frac{s}{l}$, which, inserted for X in formula 1, gives

$$F = \frac{Pl}{s} + \frac{Pl}{s}\, hyp.log.\,\frac{s}{l} = \frac{Pl}{s}\left(1 + hyp.log.\frac{s}{l}\right) \quad . \quad 2$$

The mean pressure for different pressures and expansion of steam is calculated by this formula, and given in a table farther on.

HYPERBOLIC LOGARITHMS.

§ 15. The common logarithm multiplied by 2.30258509 gives the hyperbolic logarithm, and the hyperbolic logarithm multiplied by 0.43429448 gives the common logarithm.

The following table contains the hyperbolic logarithms for numbers up to 39, which is considered sufficient for application to expansion of steam.

TABLE III.

Hyperbolic Logarithms.

No.	Logarithms.	No.	Logarithms.	No.	Logarithms.	No.	Logarithms.
1.	0.00000	4.	1.38629	7.	1.94591	10	2.30258
1.1	0.09530	4.1	1.41096	7.1	1.96006	11	2.39589
1.2	0.18213	4.2	1.43505	7.2	1.97406	12	2.48491
1.3	0.26234	4.3	1.45859	7.3	1.98787	13	2.56494
1.4	0.33646	4.4	1.48161	7.4	2.00149	14	2.63906
1.5	0.40505	4.5	1.50408	7.5	2.01490	15	2.70805
1.6	0.46998	4.6	1.52603	7.6	2.02816	16	2.77259
1.7	0.53063	4.7	1.54753	7.7	2.04115	17	2.83321
1.8	0.58776	4.8	1.56859	7.8	2.05415	18	2.89037
1.9	0.64181	4.9	1.58922	7.9	2.06690	19	2.94444
2.	0.69315	5.	1.60944	8.	2.07944	20	2.99573
2.1	0.74190	5.1	1.62922	8.1	2.09190	21	3.04452
2.2	0.78843	5.2	1.64865	8.2	2.10418	22	3.09104
2.3	0.83287	5.3	1.66770	8.3	2.11632	23	3.13549
2.4	0.87544	5.4	1.68633	8.4	2.12830	24	3.17805
2.5	0.91629	5.5	1.70475	8.5	2.14007	25	3.21888
2.6	0.95548	5.6	1.72276	8.6	2.15082	26	3.25810
2.7	0.99323	5.7	1.74046	8.7	2.16338	27	3.29584
2.8	1.02962	5.8	1.75785	8.8	2.17482	28	3.33220
2.9	1.06473	5.9	1.77495	8.9	2.18615	29	3.36730
3.	1.09861	6.	1.79175	9.	2.19722	30	3.40120
3.1	1.13140	6.1	1.80827	9.1	2.20837	31	3.43399
3.2	1.16314	6.2	1.82545	9.2	2.21932	32	3.46574
3.3	1.19594	6.3	1.84055	9.3	2.23014	33	3.49651
3.4	1.22373	6.4	1.85629	9.4	2.24085	34	3.52636
3.5	1.25276	6.5	1.87180	9.5	2.25129	35	3.55535
3.6	1.28090	6.6	1.88658	9.6	2.26191	36	3.58352
3.7	1.30834	6.7	1.90218	9.7	2.27228	37	3.61092
3.8	1.33046	6.8	1.91689	9.8	2.28255	38	3.63759
3.9	1.36099	6.9	1.93149	9.9	2.29171	39	3.66356

THERMO-DYNAMICS.

§ 16. The thermo-dynamic equivalent of heat as adopted by Joule is 772 foot-pounds of work per unit of heat.

Different authors have given different values of this equivalent—namely,

	Foot-pounds.		Foot-pounds.
Joule	772	Joule in 1843	835
Favré	750	Le Roux " 1857	832
Hirn	723	Regnault " 1871	792
Quintus	712	Violle " 1874	790

It is not necessary for the purpose of this elementary treatise to enter into an investigation of what is the true equivalent of heat, because a constant equivalent cannot be realized in the working of a steam-engine; for which reason we will here limit ourselves only to the operation of evaporating water into steam, and its transmission through a steam-engine with or without expansion.

The thermo-dynamic equivalent of heat is the ratio obtained by dividing the work in foot-pounds by the number of units of heat which performs that work.

Formula 2, § 6, gives the work of evaporation of a volume of water W, under a steam-pressure P, without expansion, or

$$K = W(\Psi - 1)144P \quad . \quad . \quad . \quad 2$$

H' = units of heat per cubic foot of steam. (See Nystrom's *Pocket-Book*, pages 400, 401.)

J = thermo-dynamic equivalent of heat, which is the work accomplished per unit of heat expended.

X = grade of expansion of steam.

The heat utilized by the evaporation of water will then be $H' \, W(\Psi - 1)$, which, divided into the work, Formula 2, gives the equivalent,

$$J = \frac{W(\Psi - 1)144P}{H' \, W(\Psi - 1)} = \frac{144P}{H'} \quad . \quad . \quad 3$$

§ 17. The column J, Table II., is calculated by this formula, and it will be seen that the equivalent varies with the steam-pressure.

When the steam is expanded, the equivalent will be increased by the hyperbolic logarithm of the expansion. When the steam is expanded say twice its volume, then $X=2$, for which the hyperbolic logarithm is 0.693, or 69 per cent. is gained by that expansion; therefore the gross equivalent realized by steam working expansively will be

$$J=\frac{144P}{H'}(1+hyp.log.X) \quad . \quad . \quad 4$$

From this formula we obtain the grade of expansion required for any value of the equivalent J—namely,

$$Hyp.log.X=\frac{JH'}{144P}-1 \quad . \quad . \quad . \quad 5$$

Example 5. How much must steam of pressure $P=100$ pounds to the square inch be expanded in order to realize Joule's equivalent $J=772$?

$$Hyp.log.X=\frac{772\times 275.52}{144\times 100}-1=13.771.$$

The number corresponding to this logarithm is 777830—that is to say, the steam must be expanded 777830 times its primitive volume in order to realize 772 foot-pounds per unit of heat; but the steam will condense to water and freeze to ice long before that expansion is reached, showing the inapplicability of Joule's equivalent to dynamics of steam.

By the new steam formulas given farther on, the thermo-dynamic equivalent is constant, 51.5 foot-pounds of work per unit of heat—that is, for full steam; and when expanded, the equivalent will be

$$J=51.5\,(1+hyp.log.X).$$

This is probably the correct thermo-dynamic equivalent of heat as realized by steam.

HORSE-POWER OF STEAM-BOILERS BY EVAPORATION.

§ 18. Heretofore it has been the custom to rate the power or steaming capacity of a boiler by the indicated horse-power transmitted by the steam-engine, and it has been found that one and the same boiler, fired under equal circumstances, but supplying steam to different engines, has produced widely different indicated horse-power, thus demonstrating that the power transmitted by the engine is not a correct measure of the real power or steaming capacity of the boiler. The question then arose, How can the power of the boiler be correctly determined independent of the working of the engine?

When a steam-user orders a boiler from a boiler-maker, it is generally specified in the contract what power the boiler must generate; but when finished and tried, the parties concerned do not agree as to what is the correct horse-power of the boiler, and law-suits have thus been instituted and unjust verdicts rendered for want of a definite rule by which to settle the question indisputably and satisfactorily to both parties.

In one case a boiler-maker contracted to furnish three boilers of 75 HP each, or in all 225 HP, for a price of $40 per horse-power, or in all $9000; but on trial, only from 100 to 130 HP was generated, according to indicator diagrams from the steam-engine.

§ 19. The steam-user, finding that power insufficient for his work, declined to pay the full price, $9000, had the boilers taken out and replaced by new ones of the requisite power, furnished by another boiler-maker.

The first boiler-maker maintained that his boilers were of the requisite power, and sued the steam-user in order to recover the full price, $9000. Several experts on steam-boiler performance were called as witnesses, and the trial of the case lasted four days, most of which time was consumed in arguing what quantity of water evaporated per hour is equivalent to one horse-power; but none of the experts appeared to understand the subject. The judge remarked that scientific evidence could not be admitted in the case, and asked if there was any reliable authority on the subject, and was answered *no*.

One expert witness stated that the boilers evaporated 100 cubic feet of water per hour under a steam-pressure of 75 pounds to the square inch, but could not state how much horse-power that evaporation would be equivalent to. No evidence was given to the fact that the boilers did not come up to 225 horse-power, and the jury rendered a verdict for the boiler-maker to receive the full pay, $9000.

The evaporation of 100 cubic feet of water per hour under a pressure of 75 pounds to the square inch is equivalent to 160 HP, and the boilers consequently did not come up to the 225 HP contracted for. Cases of this kind have frequently occurred and caused much inconvenience to the parties concerned.

The horse-power of a steam-boiler can be determined correctly by the quantity of water evaporated per unit of time independent of the working of the steam-engine, supposing that all the water is evaporated and nothing carried over in the form of foam, known as priming. A distinct line can thus be traced between the efficiencies of the power-generator and the motor.

§ 20. The horse-power given by the indicator diagrams depends much upon the construction of the engine, the regulation of the steam-valves, the grade of expansion used and the correctness of the indicator, with which the boiler-maker has nothing to do, and for which the performance of the boiler should not be held responsible.

The steam-engine may be connected with the boiler by a long, narrow and uncovered steam-pipe, in which steam may condense by radiation of heat, and the steam cylinder may be uncovered, which reduces the indicated horse-power.

§ 21. A condensing or compound engine working with a high degree of expansion indicates much more power per water evaporated than does a non-condensing engine working with full steam, which difference of power depends upon the engine-builder, and not upon the boiler-maker.

The question may arise whether the steam-pressure of the horse-power should be taken above vacuum or above the atmospheric pressure. The boiler-maker may argue that the steam generated in his boiler drives out the atmospheric pressure, and thus claim the right to be credited with the gross power of the steam supplied from his boiler.

The steam-user, on the other hand, cannot realize all that power for his work, and is therefore not willing to pay for more than value received. The boiler-maker does not undertake to remove the atmospheric pressure from the back side of the cylinder-piston, which is partly done by the engine-builder making a condensing engine, for which the power of the boiler should include only the pressure indicated by a proper steam-gauge or safety-valve, which is the pressure for estimating the power of a non-condensing engine.

§ 22. The legal horse-power of a steam-boiler fired with a given kind or quality of fuel should therefore be that passing from the boiler into the steam-pipe, with the pressure above that of the atmosphere, independent of the indicated power transmitted by the steam-engine.

When a water-owner rents out a waterfall, he only furnishes the natural effect, and does not hold himself responsible for the efficiency of the water-wheel which the miller may employ for realizing the power of that fall. It is to the interest of the miller to use the best wheel that will utilize the highest percentage of the definite natural effect of the waterfall.

So it should be also with boilers and engines. The boiler-maker furnishes a steam-boiler generating a definite natural effect of unexpanded steam, and it is to the interest of the steam-user to employ the best construction of engine in order to utilize the highest percentage of the natural effect of that steam.

The price of a steam-boiler should be rated according to the natural effect it generates with a given quality and quantity of fuel consumed per unit of time, and the boiler-maker should not be entitled to remuneration for the effect realized by the superior construction of the steam-engine, which credit is due to the engine-builder, who is paid therefor by the steam-user.

§ 23. The legal horse-power of a steam-boiler should therefore be that determined by Formula 11, § 7, with the exception that the steam-pressure p should be taken above that of the atmosphere—namely,

$$\text{HP} = \frac{Wp(V - 1)}{13748.4} \quad . \quad . \quad . \quad 1$$

The water required to be evaporated per hour for a given horse-power is

$$W = \frac{13748.4\,\text{HP}}{p(V - 1)} \quad . \quad . \quad . \quad . \quad 2$$

log. 13748.4 = 4.1382522.

W = cubic feet of water of temperature 32° Fahr. evaporated per hour.

V = steam-volume compared with that of its water at 32° Fahr. See pages 400, 401, Nystrom's *Pocket-Book.*

When the water evaporated per hour is expressed in pounds, the Formulas 1 and 2 will be

$$\text{HP} = \frac{\text{lbs.}\ p\,(V - 1)}{857721} \quad . \quad . \quad . \quad 3$$

$$\text{lbs.} = \frac{857721\,\text{HP}}{p(V - 1)} \quad . \quad . \quad . \quad . \quad 4$$

The term "legal" is used on the ground that the formulas are based upon Watt's unit of horse-power, which unit is legalized all over the civilized world, differing only slightly in different countries, to accommodate the different units of weight and measure; therefore the legalization of Watt's rule for horse-power makes the formulas in this paragraph legal.

Watt's unit for horse-power is 33000 minute-foot-pounds, which is the same as 550 second-foot-pounds, the standard upon which the formulas are based.

Example 3. What is the horse-power of a boiler evaporating ℔s. = 640 pounds of water per hour of temperature $t = 80°$, to steam of $p = 80$ pounds to the square inch?

$$\text{HP} = \frac{640 \times 80(280.5 - 1)}{857721} = 166.84.$$

Correction for temperature of feed-water 80° will be $0.9577 \times 166.84 = 159.78$, the horse-power required.

Example 1. A steam-boiler evaporating $W = 64$ cubic feet of water per hour, under a pressure of $p = 85$ pounds to the square inch above that of the atmosphere, the temperature of the feed-water being $t = 120°$ Fahr. Required the legal horse-power of the boiler?

$$\text{HP} = \frac{64 \times 85 \times 266.8}{13748.4} = 105.57.$$

Correction for temperature 120° of the feed-water will be, see Table I., page 22.

$\text{HP} = 0.9199 \times 105.57 = 97.113$, the legal horse-power required.

Example 2. How much feed-water of $t = 90°$ must be evaporated per hour under a pressure of $p = 65$ pounds to the square inch above that of the atmosphere in order to generate a legal horse-power $\text{HP} = 360$ horses of the boiler?

$$W = \frac{13748.4 \times 360}{65 \times 327.08} = 232.8 \text{ cu. ft.}$$

Correction for temperature $t = 90°$ $W = 232.8 : 0.9486 = 245.43$ cubic feet, the quantity of water required.

The following Table IV. is calculated from the above formulas, giving the quantity of water expressed in cubic feet, cubic inches or pounds required to be evaporated per hour per horse-power, and also the horse-power per cubic foot of water evaporated per hour under different pressures.

TABLE IV.

Legal Horse-power of Steam-boilers per Rate of Evaporation of Water to Steam.

Steam pressure ab. atm.	Water evaporated per hour per horse-power.			Horse-power per cub. ft.	Work. ft.-lbs. per unit of heat.
	Cubic feet.	Cubic in.	Pounds.		
p	*W*	*w*	*lbs.*	HP	*J*
5	2.2562	3898.8	140.76	0.4433	12.225
10	1.3983	2416.2	87.235	0.7150	19.616
15	1.1106	1919.0	69.284	0.9005	24.701
20	0.9654	1668.1	60.226	1.0358	28.380
25	0.9770	1515.4	54.711	1.1403	31.145
30	0.8176	1411.9	51.010	1.2231	33.433
35	0.7743	1338.1	48.308	1.2914	35.171
40	0.7412	1280.9	46.244	1.3490	36.683
45	0.7150	1235.5	44.605	1.3986	37.988
50	0.6935	1198.3	43.264	1.4420	39.124
55	0.6755	1167.2	42.140	1.4804	40.118
60	0.6600	1140.6	41.180	1.5150	41.012
65	0.6467	1117.5	40.345	1.5463	41.819
70	0.6349	1097.1	39.607	1.5750	42.551
75	0.6243	1078.9	38.951	1.6016	43.221
80	0.6149	1062.5	38.360	1.6263	43.854
85	0.6062	1046.6	37.822	1.6495	44.425
90	0.5983	1033.9	37.328	1.6713	45.011
95	0.5910	1021.3	36.873	1.6919	45.533
100	0.5847	1009.6	36.451	1.7115	46.027
105	0.5779	998.67	36.056	1.7303	46.495
110	0.5720	988.44	35,686	1.7482	46.949
115	0.5664	978.75	35.337	1.7655	47.390
120	0.5611	969.62	35.007	1.7822	47.812
125	0.5561	960.96	34.694	1.7982	48.213
130	0.5513	952.65	34.394	1.8073	48.604
135	0.5468	945.00	34.111	1.8288	49.737

HORSE-POWER OF STEAM-BOILERS BY FIRE-GRATE AND HEATING SURFACE.

§ 24. The evaporating capacity of a steam-boiler fired with a given kind or quality of fuel depends upon the extent of area of fire-grate and heating surface.

$\boxminus$ = area of fire-grate in square feet.

$\square$ = heating surface in square feet.

W = cubic feet of water of temperature 32° Fahr. evaporated per hour.

In ordinary steam-boilers the average evaporation with natural draft is

$$W = 0.4\sqrt{\boxminus\,\square}, \quad . \quad . \quad . \quad . \quad 1$$

under the condition that the heating surface should be between 18 and 36 times the area of the fire-grate.

This water, multiplied by the steam-volume, gives the space generated per hour by the steam, or

$$S = 0.4\,(\mathfrak{V} - 1)\sqrt{\boxminus\,\square}, \quad . \quad . \quad . \quad 2$$

S = cubic feet of steam generated per hour.

This space, multiplied by the steam-pressure per square foot 144 P, gives the work accomplished by the steam per hour.

$$K = 0.4\,(\mathfrak{V} - 1)\sqrt{\boxminus\,\square}\;144\,P \quad . \quad . \quad . \quad 3$$

$$K = 57.6\,P(\mathfrak{V} - 1)\sqrt{\boxminus\,\square} \quad . \quad . \quad . \quad 4$$

This work, divided by 33,000 pounds times 60 minutes = 1,980,000, gives the horse-power of the boiler expressed by area of fire-grate and heating surface.

$$\text{HP} = \frac{57.6\,P(\mathfrak{V} - 1)\sqrt{\boxminus\,\square}}{1980000} \quad . \quad . \quad . \quad 5$$

$$\text{HP} = \frac{P(\mathfrak{V} - 1)\sqrt{\boxminus\,\square}}{34400} \quad . \quad . \quad . \quad 6$$

This formula gives the gross effect or horse-power of the steam above vacuum; but for the practical rating of the power of a steam-boiler, the pressure should be taken above that of the atmosphere, or

$$\text{HP} = \frac{p\,(\mathfrak{V} - 1)\sqrt{\boxminus\,\square}}{34400} \quad . \quad . \quad . \quad 7$$

§ 25. Although this formula gives the average horse-power of a steam-boiler, it cannot be termed legal because the evaporation is the real power of the boiler, which depends upon the firing, circulation of water and other variable circumstances not included in the formula.

The volume $(\psi - 1)$ varies inversely with the pressure, and the product $p\,(\psi - 1)$ varies nearly as the cube root of the pressure p, for which we may practically place $p\,(\psi - 1) = 5000\sqrt[3]{p}$, and the horse-power of the boiler will be for a chimney 50 feet high,

$$\text{HP} = \frac{\sqrt[3]{p}\sqrt{\boxminus\,\square}}{6.88} \quad . \quad . \quad . \quad . \quad 8$$

See Table of Corrections for Height of Chimney.

This formula should not be used for less pressure than $p = 15$ pounds to the square inch.

Example 8. Required the horse-power of a steam-boiler with fire-grate $\boxminus = 130$ square feet, and heating-surface $\square = 3372$ square feet, carrying a steam-pressure of $p = 50$ pounds to the square inch above that of the atmosphere?

$$\text{Formula 42.} \quad \text{HP} = \frac{\sqrt[3]{50}\sqrt{130 \times 3372}}{6.88} = 354.53 \text{ horse-power.}$$

This is the horse-power the boiler would generate without expanding the steam.

Example 1. Required the quantity of water evaporated per hour by the fire-grate $\boxminus = 130$, and heating-surface $\square = 3372$ square feet?

$$\text{Formula 1.} \quad W = 0.4\sqrt{130 \times 3372} = 264.8 \text{ cubic feet.}$$

The efficiencies of steam-boilers can readily be compared with these formulas.

When the steam is expanded in the engine, the power derived from the boiler may practically be estimated as follows:

$$\text{HP} = \frac{\sqrt[3]{p}\sqrt{\boxminus\,\square}}{6.4}(1 + hyp.log.X) \quad . \quad . \quad 9$$

This formula should not be taken for the horse-power of the steam-boiler, but for that transmitted by the engine.

In the preceding example we have the horse-power $\text{HP} = 354.53$ when working with full steam; but if the steam is expanded say $X = 3$ times, we have $hyp.log.3 = 1.0986$ and $2.0986 \times 354.53 = 744$

horse-power, of which 70 per cent. may be transmitted through the engine, or

$$\text{HP} = 744 \times 0.7 = 520.8 \text{ horse-power},$$

which would probably be indicated by diagrams.

It is supposed in the preceding examples that the temperature of the feed-water is 32°. For any other temperature up to 212°, use the correction in the following Table V., corresponding to the actual temperature of the feed-water, as follows:

R = correction in the Table V.

$$W = 0.4R\sqrt{\equiv \square} \quad . \quad . \quad . \quad . \quad 1$$

$$\text{HP} = \frac{pR(\sqrt[\prime]{} - 1)\sqrt{\equiv \square}}{34400} \quad . \quad . \quad . \quad 7$$

$$\text{HP} = \frac{R\sqrt[8]{p}\sqrt{\equiv \square}}{6.88} \quad . \quad . \quad . \quad . \quad 8$$

TABLE V.

Gain of Power or Water evaporated by heating the Feed-water from 32° to t.

Temp. t.	Gain R.	Logarithm.	Temp. t.	Gain R.	Logarithm.
40	1.0068	0.0029432	130	1.0983	0.0407210
50	1.0151	0.0065088	140	1.1111	0.0457531
60	1.0245	0.0105120	150	1.1221	0.0500316
70	1.0340	0.0145205	160	1.1344	0.0547662
80	1.0441	0.0187421	170	1.1469	0.0595256
90	1.0542	0.0229230	180	1.1594	0.0642333
100	1.0647	0.0272273	190	1.1725	0.0691129
110	1.0757	0.0316912	200	1.1859	0.0740481
120	1.0870	0.0362295	212	1.2023	0.0800128

The horse-power given by Formulas 8, multiplied by the correction for height of chimney, Table VII., gives the horse-power which may be expected from the boiler.

The following Table VI. gives the horse-power of boilers per square foot of grate-surface for different proportions of heating-surface, when the steam is worked without expansion through a non-condensing engine, and the temperature of the feed-water is 32°.

TABLE VI.

Horse-power per Square Foot of Fire-grate for Chimney 50 Feet High.

Steam pressure.	Proportion of fire-grate and heating surface.					
	□ = 16 ▤	□ = 20 ▤	□ = 25 ▤	□ = 30 ▤	□ = 35 ▤	□ = 40 ▤
p	HP	HP	HP	HP	HP	HP
25	1.7	1.91	2.14	2.33	2.52	2.63
30	1.81	2.02	2.27	2.48	2.67	2.8
35	1.91	2.13	2.38	2.61	2.81	2.95
40	2.	2.23	2.49	2.73	2.94	3.08
45	2.08	2.32	2.59	2.84	3.06	3.2
50	2.15	2.4	2.68	2.94	3.17	3.32
55	2.22	2.48	2.77	3.03	3.28	3.42
60	2.29	2.55	2.85	3.12	3.37	3.52
65	2.35	3.62	2.93	3.2	3.46	3.62
70	2.4	3.67	2.99	3.27	3.53	3.7
75	2.46	2.74	3.07	3.36	3.63	3.8
80	2.51	2.81	3.13	3.43	3.71	3.88
85	2.57	2.87	3.2	3.51	3.79	3.96
90	2.62	2.92	3.26	3.57	3.85	4.04
95	2.66	2.97	3.32	3.63	3.93	4.11
100	2.71	3.02	3.38	3.7	4.	4.19
110	2.8	3.12	3.49	3.82	4.13	4.32
120	2.88	3.21	3.59	3.93	4.24	4.44
130	2.96	3.3	3.68	4.04	4.36	4.57
140	3.03	3.38	3.78	4.14	4.47	4.68
150	3.11	3.46	3.87	4.23	4.57	4.79
160	3.17	3.54	3.95	4.33	4.67	4.89
170	3.23	3.62	4.03	4.42	4.77	4.99
180	3.3	3.68	4.11	4.5	4.86	5.09
190	3.36	3.74	4.21	4.58	4.94	5.18
200	3.41	3.81	4.26	4.64	5.03	5.27
210	3.47	3.88	4.32	4.74	5.11	5.36
220	3.52	3.93	4.39	4.81	5.2	5.44
230	3.58	3.99	4.46	4.88	5.28	5.52
240	3.63	4.05	4.52	4.95	5.35	5.6
250	3.68	4.11	4.59	5.03	5.42	5.68
260	3.72	4.16	4.65	5.09	5.49	5.75
270	3.77	4.21	4.70	5.16	5.57	5.82
280	3.82	4.26	4.76	5.22	5.63	5.9
290	3.87	4.31	4.82	5.28	5.69	5.96
300	3.9	4.36	4.87	5.34	5.75	6.03

TABLE VII.

Correction of Horse-power per Square Foot of Grate for Different Heights of Chimneys in Feet.

Height Chimney.	Correction.	Height Chimney.	Correction.	Height Chimney.	Correction.	Height Chimney.	Correction.
feet.	r.	feet.	r.	feet.	r.	feet.	r.
10	0.5	75	1.20	180	1.78	310	2.27
15	0.59	80	1.23	190	1.82	320	2.30
20	0.67	85	1.27	200	1.86	330	2.33
25	0.74	90	1.30	210	1.90	340	2.36
30	0.8	95	1.33	220	1.94	350	2.40
35	0.85	100	1.36	230	1.98	360	2.43
40	0.91	110	1.42	240	2.02	370	2.46
45	0.96	120	1.48	250	2.06	380	2.49
50	1.00	130	1.53	260	2.10	390	2.52
55	1.04	140	1.58	270	2.13	400	2.55
60	1.08	150	1.63	280	2.16	410	2.57
65	1.12	160	1.68	290	2.20	420	2.60
70	1.16	170	1.73	300	2.23	430	2.63

Allowance is made in the above table for radiation or conduction of heat from the gases through the walls of the chimney.

TABLE VIII.

Consumption of Coal in Pounds per Hour per Square Foot of Grate, for Different Heights of Chimney.

Height Chimney.	Consumpt. coal.	Height Chimney.	Consumpt. coal.	Height Chimney.	Consumpt. coal.	Height Chimney.	Consumpt. coal.
10	7.00	75	16.8	180	25.	310	31.8
15	8.25	80	17.2	190	25.5	320	32.2
20	9.4	85	17.8	200	26.	330	32.7
25	10.4	90	18.2	210	26.5	340	33.1
30	11.2	95	18.6	220	27.2	350	33.6
35	12.	100	19.	230	27.7	360	34.
40	12.8	110	19.9	240	28.3	370	34.4
45	13.4	120	20.7	250	28.9	380	34.9
50	14.	130	21.4	260	29.4	390	35.3
55	14.6	140	22.1	270	29.8	400	35.7
60	15.1	150	22.9	280	30.3	410	36.
65	15.7	160	23.5	290	30.8	420	36.4
70	16.2	170	24.2	300	31.2	430	36.8

It is not expected that this gives the correct consumption of coal, which depends much upon the kind of coal and manner of firing, but it gives the proportionate consumption to the height of the chimney. See horse-power of chimney, Table XXIX., page 123.

CHIMNEYS.

§ 26. The proportion of a chimney to the horse-power of the steam generated and consumption of fuel on the fire-grate is a very difficult problem to solve theoretically. It is certain, however, that the horse-power of a chimney, as well as the consumption of fuel on the fire-grate, is directly as the section area and square root of the height of the chimney.

The term "horse-power" in this connection means the power of draft in a chimney required for the combustion generating heat for evaporation of water to steam of a given horse-power.

The following formulas are derived from both theory and practice, and the horse-power is that generated by full steam without expansion :

HP = horse-power of chimney.

$\boxminus$ = area of fire-grate in square feet.

A = section area of chimney in square feet.

H = height of chimney in feet when $A = 0.16 \boxminus$.

C = pounds of coal consumed per hour on the fire-grate.

r = coefficient for correction in the preceding Table VII.

Horse-power,	$\text{HP} = 10\,A\,r$	1
Consumption of fuel,	$C = 14 \boxminus r$	2
Area of chimney,	$A = \dfrac{\text{HP}}{10\,r}$	3
Area of grate,	$\boxminus = \dfrac{C}{14\,r}$	4
Correction,	$r = \dfrac{\text{HP}}{10\,A}$	5
Correction,	$r = \dfrac{C}{14 \boxminus}$	6
Correction,	$r = \dfrac{\sqrt{H} + \sqrt[3]{H}}{10.755}$	7

Example 1. Required the horse-power of a chimney $H = 80$ feet high above grate and $A = 4$ square feet cross-section?

Correction for 80 feet = 1.23.

$$\text{HP} = 10 \times 4 \times 1.23 = 49.2 \text{ horse-power.}$$

Example 2. How much coal will be consumed per hour on a fire-grate $⊟ = 150$ square feet connected with a chimney $H = 60$ feet high?

Correction for 60 feet is 1.09.

$$C = 14 \times 150 \times 1.09 = 2289 \text{ pounds.}$$

Example 6. What height of chimney is required for a draft consuming $C = 1216$ pounds of coal per hour on a grate $⊟ = 64$ square feet?

Correction, $$r = \frac{1216}{14 \times 64} = 1.357.$$

The height of chimney in the table corresponding to this correction is $H = 100$ feet.

Example 5. A chimney is to be constructed for a boiler having a grate surface of $⊟ = 48$ square feet. The section area of the chimney is made $A = 0.16 ⊟ = 0.16 \times 48 = 7.68$ square feet. How high must the chimney be that the draft will generate $HP = 192$ horse-power?

Correction, $$r = \frac{192}{10 \times 7.68} = 2.5. \quad \text{Height } H = 390 \text{ feet.}$$

The smoke-stacks for steamboats are generally made cylindrical or parallel—that is, of equal section from boiler to top; but brick chimneys for factories are generally made taper, with about 45 per cent. more section area at the bottom than at the top. The area A in the preceding formulas and examples should be that at the top of the chimney.

POWER OF COMBUSTION.

§ 27. On account of the physical constitution of heat not being well understood, an intelligent explanation of dynamics of combustion cannot be given.

Combustion is the operation of combining oxygen with fuel, which generates heat; and the more rapidly that combination is performed, the higher will be the temperature of the heat.

The chemical combination of oxygen with a definite weight of fuel generates a definite quantity of heat, which is convertible into work, or the product of the three simple physical elements *force, velocity* and *time*, represented by $F\,V\,T$. Of this work, the thermo-dynamic equivalents may be represented as follows:

F = force, which is convertible into temperature of the heat.

V = velocity, or rate of combustion, which is proportioned to the area of the fire-grate.

$F\,V$ = power, the act of combustion, or combination of oxygen and fuel.

$V\,T$ = space, or the volume occupied by the heat.

$F\,V\,T$ = work, which represents the quantity of heat generated by the combustion in the time T.

For a definite quantity of heat generated in a long time T the power $F\,V$ must be small, and for a short time T the power $F\,V$ must be larger; but for a constant power $F\,V$ either one of the elements F and V may vary at the expense of the other.

§ 29. For a definite quantity of fuel consumed per unit of time on different extent of grate-surface, the temperature of combustion should be inversely as the grate-surface—that is to say, a forced draft should generate a higher temperature of the fire than would be attained by natural draft.

The combustion per unit of time is as the square root of the pressure of the air.

The fuels generally used for generating heat are *carbon*, *hydrogen* and *sulphur*, of which only carbon, which is the predominant fuel used in steam-boilers, will herein be considered. Carbon forms two compounds with oxygen—namely, carbonic oxide $C\,O$, and carbonic acid $C\,O_2$, the equivalent of carbon being 6, and that of oxygen 8—that is to say, 6 pounds of carbon united with 8 pounds of oxygen forms carbonic oxide, which is a transparent, colorless gas which when ignited will burn with a faint flame, taking up another atom of oxygen, and forms carbonic acid, composed of 6 pounds of carbon and $8 \times 2 = 16$ pounds of oxygen.

One pound of carbon combined with $16 : 6 = 2\frac{2}{3}$ pounds of oxygen forms $3\frac{2}{3}$ pounds of carbonic acid, which is complete combustion of the carbon.

AIR FOR COMBUSTION.

§ 30. The oxygen required for combustion is supplied from atmospheric air, which is a mechanical mixture of

23 weights of oxygen to 77 of nitrogen in 100 weights of air.

21 volumes of oxygen to 79 of nitrogen in 100 volumes of air.

One cubic foot of dry atmospheric air, of temperature 60° Fahr. and under a pressure of 30 inches of mercury, weighs 532 grains, or 0.076 of a pound, and 13.158 cubic feet weighs one pound.

One pound of air contains 0.23 pounds of oxygen, and $13.158 : 0.23 = 57.21$ cubic feet of air to make one pound of oxygen.

The combustion of one pound of carbon requires $2\frac{2}{3}$ pounds of oxygen; therefore $57.21 \times 2\frac{2}{3} = 152.56$ cubic feet of dry air, of temperature 60°, is required for the complete combustion of one pound of carbon.

Carbonic oxide requires $57.21 \times 1\frac{1}{3} = 76.28$ cubic feet of air per pound of carbon consumed.

Different temperatures of the air require different volumes for the combustion of one pound of carbon, as shown in the accompanying Table IX.

TABLE IX.

Properties of Air for Combustion.

Temp. of air. Fahr.	Volume of air. 1 at 32°.	Weight per cub. foot. ℔	Cubic feet of air required for			
			1 lb. of air.	1 lb. of oxygen.	Comb. 1 lb. carbon. carb. acid.	carb. oxide.
10	0.9554	0.08414	11.885	51.674	137.804	68.902
20	0.9756	0.08236	12.142	52.792	140.778	70.389
32	1.0000	0.08023	12.464	54.191	144.510	72.255
40	1.0162	0.07886	12.681	55.135	147.026	73.513
50	1.0365	0.07718	12.957	56.335	150.226	75.113
60	1.0567	0.07600	13.158	57.209	152.556	76.278
70	1.0760	0.07453	13.417	58.335	155.560	77.780
80	1.0973	0.07311	13.678	59.470	158.586	79.293
90	1.1176	0.07146	13.994	60.843	162.248	81.124
100	1.1378	0.07051	14.182	61.661	164.430	82.215
110	1.1581	0.06928	14.434	62.756	167.348	83.674
120	1.1784	0.06808	14.688	63.861	170.296	85.148

TEMPERATURE OF DRAFT.

§ 31. In comparative experiments on evaporation or steaming capacities of boilers supplied with air of widely different temperatures, various opinions have been advanced as to what would be the proper allowance for temperature of the air.

When the air of different temperatures enters the furnace under constant pressure or natural draft, what is gained by the warmer air is lost by less oxygen per volume.

In a cold atmosphere there is better draft in the chimney than in warmer air; but when the air is supplied and heated under pressure, as in a blast-furnace, then there is an advantage in the combustion by the hot air.

In a cold atmosphere more heat will no doubt be radiated from the boiler and steam-pipe, but the generation of heat in the furnace and steam in the boiler will not be materially diminished, although the cold air enters the fire with less velocity than does warmer air.

HEAT OF COMBUSTION.

§ 31. The heat of combustion means the quantity of heat generated by the burning of a given weight of fuel, and which is a distinct quantity from that of the temperature of the fire.

The English unit of heat is that required to elevate the temperature of one pound of water from 39° to 40° Fahr. The experiments of Regnault show that the elevation of the temperature of one pound of water from 32° to 212° or 180° requires 180.9 units of heat—that is to say, for higher temperatures than 39° to 40° it requires a little more than one unit of heat to elevate the temperature of one pound of water one degree; but the difference is so small that in practice we may consider one unit of heat as standard for elevating the temperature of one pound of water one degree at all temperatures below that of the boiling point.

The experiments of Favre and Silberman show that the combustion of one pound of carbon to $2\frac{1}{3}$ pounds of carbonic oxide generates 4400 units of heat, and to $2\frac{2}{3}$ pounds of carbonic acid 14,500 units of heat. That is to say, the acid generates 14,500 : 4400 = 3.27 times more heat than does the oxide, showing the importance of burning the fuel completely to acid. If it requires, say, 150 cubic feet of air for burning one pound of carbon to acid, it requires only 75 cubic feet for the burning to oxide. Now, if the supply of air is between 150 and 75 cubic feet, both the gases will be formed and mechanically mixed, but not chemically combined, in the combustion chamber.

Suppose 120 cubic feet of air is supplied per pound of carbon consumed, what will be the proportion of oxide and acid formed? and how many units of heat will be generated per pound of carbon consumed?

Assuming the temperature of the air to be 60°, it requires 57.21 cubic feet to make one pound of oxygen, and 120 : 57.21 = 2.0975, say two pounds of oxygen.

$$\text{Carbonic oxide} = \frac{56 - 21 \times 2}{12} = 1.1666 \text{ pounds.}$$

$$\text{Carbonic acid} = \frac{33 \times 2 - 44}{12} = 1.8333 \text{ pounds.}$$

One pound of carbonic oxide generates 1650 units of heat. One pound of carbonic acid generates 3960 units of heat. Then $1650 \times 1.1666 + 3960 \times 1.8333 = 9184.75$ units of heat generated by the combustion of one pound of carbon with the oxygen of 120 cubic feet of air. With 30 cubic feet, or 25 per cent., more air the carbon would have been consumed to acid, and generated 14,500, or nearly 58 per cent. more heat. This shows the importance of supplying a sufficient quantity of air to the furnace for the complete combustion of the carbon to carbonic acid.

§ 32. FORMULAS FOR HEAT AND COMBUSTION.

CO = pounds of carbonic oxide, } formed by combustion.
CO_2 = pounds of carbonic acid, } formed by combustion.
C = pounds of carbon consumed by
O = pound of oxygen.
h = units of heat generated by the combustion.

$$CO = \frac{56\,C - 21\,O}{12} \quad . \quad . \quad . \quad . \quad 1$$

$$CO_2 = \frac{33\,O - 44\,C}{12} \quad . \quad . \quad . \quad . \quad 2$$

$$h = 3960\,(CO_2) + 1650\,(CO) \quad . \quad . \quad . \quad 3$$

The following Table X. is calculated by the above formulas, making $C = 1$ pound of carbon. The first column contains the oxygen supplied for the combustion of one pound of carbon, and the second column the cubic feet of air containing the oxygen in the first column:

TABLE X.

Operation of Incomplete Combustion of Carbon.

Per lb. of Carbon.		Carbonic Oxide.		Carbonic Acid.		Total units of heat.	Percentage of heat lost.
Oxygen lbs.	Air 60° cub. feet.	CO lbs.	Units of heat.	CO_2 lbs.	Units of heat.		
$1.\frac{1}{3}$	76.278	$2\frac{1}{3}$	4400	0	0	4400	69.65
1.4	80.092	2.2222	3666.6	0.1833	726.0	4713.6	67.02
1.5	85.813	2.0416	3368.6	0.4583	1813.4	5182.0	64.26
1.6	91.534	1.8666	3080.0	0.7333	2904.0	5984.0	58.73
1.7	97.265	1.6916	2791.3	0.9258	3666.1	6457.4	55.47
1.8	102.99	1.5166	2502.5	1.2833	5082.0	7584.5	47.69
1.9	108.71	1.3416	2213.8	1.5583	6169.7	8382.5	42.19
2.0	114.42	1.1666	1925.0	1.8333	7260.0	9185.0	36.66
2.1	120.14	0.9916	1636.3	2.1083	8349.0	9985.3	31.14
2.2	125.86	0.8166	1347.5	2.3833	9438.0	10785	25.62
2.3	131.58	0.6416	1058.8	2.6583	10527	11586	20.10
2.4	137.30	0.4666	770.0	2.9333	11616	12386	14.58
2.5	143.02	0.2916	481.3	3.2083	12705	13186	9.06
$2.\frac{2}{3}$	152.55	0.0000	0000	3.6666	14500	14500	0.00

Suppose 120.14 cubic feet of air is supplied per pound of carbon consumed, the results will be as in the table—namely.

Carbonic oxide $CO = 0.9916$ lbs. of 1636.3 units of heat.
Carbonic acid $CO_2 = 2.1083$ lbs. of 8349. units of heat.

Products of combustion = 3.0999 lbs. of 9985.3 units of heat.

The loss by incomplete combustion is 31.14 per cent., as shown in the last column of the table.

This table shows the operation of incomplete combustion with a different supply of air per pound of carbon consumed. For instance, if 114.42 cubic feet of air is supplied per pound of carbon consumed, it will generate 1.16 pounds of CO of 1925 units of heat and 1.83 pounds of CO_2 of 7260 units of heat; in all 9185 units of heat, with 36.6 per cent. loss of that if 152.55 cubic feet of air had been supplied.

When less air is supplied than is required for forming carbonic acid, the produce of combustion will form smoke with unconsumed particles of carbon; and when more air is supplied than is required for forming carbonic acid, the excess will be heated by the products of combustion, which heat is thus lost and carried up through the chimney.

FUEL.

§ 33. The fuels generally used in steam-boilers for combustion to generate heat are *wood, charcoal, peat, mineral coal* and *coke*, none of which is pure carbon, as heretofore assumed in the operation of combustion, but contains various proportions of carbon, hydrogen, oxygen and involatilizable matter forming ash. The hydrogen in the fuel, combined with oxygen by combustion, generates about four times as much heat per weight of hydrogen consumed as does an equal weight of carbon. The combustion of one pound of hydrogen by 8 pounds of oxygen forms steam and generates 62032 units of heat; therefore, if one pound of fuel contains, say 0.9 of a pound of carbon and 0.1 of a pound of hydrogen, the heat generated by the combustion will be

Hydrogen, $62032 \times 0.1 = 6203.2$ units of heat.
Carbon, $14500 \times 0.9 = 13050$ " "
Total, $= 19253.2$ units of heat.

When the fuel contains only carbon and hydrogen, the following forms for combustion give the units of heat generated:

C' = fraction of carbon } in one pound of fuel.
H' = " hydrogen }

O = pounds of oxygen required for the complete combustion per pound of fuel.

$$O = 8H' + 2\tfrac{2}{3}C = 2\tfrac{2}{3}(3H' + C') \quad . \quad . \quad 1$$

A = cubic feet of air at 60° required for the combustion of one pound of fuel.

$$A = 153(3H' + C') \quad . \quad . \quad . \quad 2$$

The units of heat generated per pound of fuel consumed will be

$$h = 62032H' + 14500C' \quad . \quad . \quad . \quad 3$$

MOISTURE IN FUEL.

§ 34. When a fuel contains both oxygen and hydrogen partly combined in the form of water or moisture, that part of the fuel will be inert in the generation of heat. One-eighth of the oxygen will be equal to the inert part of the hydrogen, so that the heat generated by the hydrogen in the fuel will be $h = 62030\left(H' - \frac{O'}{8}\right)$. 4

Heat by the carbon, $h = 14500C'$ . . . 5

The sum of these two formulas will be the heat generated by the fuel when sufficient oxygen is supplied for its combustion—namely,

$$h = 14500\left(C' + 4.28(H' - \frac{O'}{8}\right) \quad . \quad . \quad 6$$

C', H' and O' are fractions in one pound of the fuel.

The weight of oxygen required for this combustion will be

$$O = 2\tfrac{1}{3}\left[C' + 3\left(H' - \frac{O'}{8}\right)\right] \quad . \quad . \quad . \quad 7$$

The cubic feet of air of 60° required for this oxygen is

$$A = 153\left[C' + 3\left(H' - \frac{O'}{8}\right)\right] \quad . \quad . \quad . \quad 8$$

UNCOMBINED OXYGEN AND HYDROGEN IN FUEL.

§ 35. When the oxygen and hydrogen in a fuel are not chemically combined, their combination by combustion will generate heat, and the oxygen required for the combustion of the C' and H' will be diminished by O'.

When a fuel contains the three combustibles carbon, hydrogen and sulphur, the heat generated by its complete combustion will be

$$h = 14500C' + 62030H' + 4032S' \quad . \quad . \quad 9$$

The proportion of ingredients in fuel varies very much, even in the same kind of fuel like mineral coal, for which analyses and experiments must be made with each fuel to determine its correct heating power.

The following Table XI. gives the average proportion and property of different fuels, compiled from analyses and experiments by the most reliable authors.

TABLE XI.

Proportions of Ingredients in, and Heat Generated by, the Combustion of One Pound of Fuel.

Fuels.	Ingredients in One Pound of Fuel. Combustibles.			Non-combustibles.			Per pound of fuel.			lbs. of fuel per cub. ft. of water evaporated.
	Carbon.	Hydrogen.	Sulphur.	Nitrogen.	Oxygen.	Ash.	Air.	Heat.	Water evap.	
	C'	H'	S'	N'	O'		Cu. ft.	h.	lbs.	
Pure Carbon	1	......	......	......	......	......	153	14500	12.4	5.03
Hydrogen	......	1	......	......	......	......	459	62032	53.	1.18
Sulphur	......	......	1	......	......	......	114.4	4032	3.44	18.2
Peat, dry	0.56	0.06	......	......	.23	0.15	100	9984	8.42	7.4
Woods, Oak	0.48	0.06	......	......	0.41	0.05	78.4	7580	6.47	9.67
" White Pine	0.49	0.08	......	......	0.39	0.04	88.7	8966	7.65	8.17
" Birch	0.48	0.07	......	......	0.40	0.05	82.6	8300	7.07	8.84
Charcoal, Oak	0.88	0.03	......	......	0.06	0.03	144	13760	11.7	5.34
" Pine	0.72	0.06	......	0.04	0.15	0.03	138.5	12921	11.	5.67
" Maple	0.70	0.05	......	0.05	0.17	0.03	121	13411	11.45	5.67
Bituminous Coal	0.84	0.05	0.015	0.012	0.03	0.05	147	14780	12.62	4.94
Anthracite Coal	0.88	......	0.01	......	......	0.06	135	12760	10.9	5.73
Coke	0.87	0.02	0.02	0.008	0.002	0.06	142	13865	11.85	5.27
C burning to CO	1	......	......	......	......	......	76.5	4400	3.76	16.6
CO burning to CO_2	0.4286	......	......	......	0.5714	......	76.5	10100	8.63	7.25
Alcohol	0.520	0.137	......	......	0.343	......	122.75	12339	10.55	5.93
Tallow	0.79	0.117	......	......	0.093	......	169	15550	13.27	4.7
Bees' Wax, White	0.815	0.139	......	......	0.045	......	186	18900	16.12	3.88

The pounds of water evaporated per pound of coal, as given in the table, is equal to the units of heat per pound of steam, of pressure $p = 50$ lbs. to the square inch above that of the atmosphere $= 1172.8$ units, divided into the units of heat generated by the combustion of one pound of coal.

The units of heat per pound of steam of any pressure is

$$h = 1082 + 0.305\,T \quad . \quad . \quad . \quad 10$$

This is the heat required to elevate the temperature of one pound of water from 32° Fahr. to boiling-point and evaporate it to steam of temperature T. See table, pages 400, 401, Nystrom's *Pocket-Book*.

When the feed-water is of higher temperature, a reduction is required as follows:

w = pounds of water heated from temperature t and evaporated to steam of temperature T per pound of fuel consumed.

h' = units of heat of combustion of one pound of coal available in evaporation.

$$h' = w(1114 + 0.305\,T - t) \quad . \quad . \quad . \quad 11$$

This is the proper formula for comparing the evaporative quality of different fuels consumed in similar boilers; and when similar fuels are used in different kinds of boilers, this formula gives the relative efficiency of the boilers.

Example. Two different kinds of fuel A and B are experimented with in one or similar boilers.

One pound of the fuel A evaporates $w = 7.5$ lbs. of water from $t = 96°$ to steam of $T = 297.84°$.

One pound of the fuel B evaporates $w = 9$ lbs. of water from $t = 115°$ to steam of $T = 311.86°$.

Required the available units of heat per pound of each fuel, and their relative steaming quality?

A. $\quad h' = 7.5(1114 + 0.305 \times 297.84° - 96°) = 8203.3$ units of heat.

B. $\quad h' = 9(1114 + 0.305 \times 311.86° - 115°) = 11917$ units of heat.

Relative quality, $\dfrac{B}{A} = \dfrac{11917}{8203.3} = 1.4527.$

The fuel B is 45½ per cent. better than the fuel A.

It is supposed that the firing and draft to the grate and other circumstances are alike in both experiments.

Example 11. Two different kinds of boilers C and D are fired with the same kind of fuel. The boiler C evaporates, per pound of coal consumed, $w = 6$ lbs. of water from $t = 60°$ to steam of $T = 393.94°$.

The boiler D evaporates, per pound of fuel consumed, $w = 8$ lbs. of water from $t = 85°$ to steam of $T = 320.1°$.

Required the relative qualities of the two boilers?

C. $\quad h' = 6(1114 + 0.305 \times 393.94° - 60°) = 7044.9$ units of heat.

D. $\quad h' = 8(1114 + 0.305 \times 320.1° - 85°) = 9013.04$ units of heat.

Relative quality of boilers, $\dfrac{D}{C} = \dfrac{9013.04}{7044.9} = 1.2794.$

The boiler D is nearly 28 per cent. better than the boiler C.

QUALITY OF BOILERS AND FUEL COMPARED WITH A STANDARD MEASURE.

§ 36. The most simple and correct way of comparing the quality or economy of different boilers fired with the same kind of fuel, or of different kinds of fuel consumed per hour in the same kind of boilers, is to compare the units of heat realized by evaporation with the total units of heat 14500 due to the combustion of one pound of carbon to carbonic acid.

In the preceding four examples *A*, *B*, *C* and *D* we have the relative economy as follows:

$$A=\frac{8203.3}{14500}=0.56575, \text{ or } 56\tfrac{1}{2} \text{ per cent.}$$

$$B=\frac{11917}{14500}=0.82186, \text{ or } 82 \text{ per cent.}$$

$$C=\frac{7045}{14500}=0.4858, \text{ or } 48\tfrac{1}{2} \text{ per cent.}$$

$$D=\frac{9013}{14500}=0.6216, \text{ or } 62 \text{ per cent.}$$

Logarithm, 14500 = 4.1613680.

The fuel *B* gave the best result, and the boiler *C* the poorest; but the question now arises whether or how much of the economy is due to the fuel or to the boiler.

The percentage of carbon in a fuel ought to determine its quality, but it is well known that different kinds of fuel of equal proportions of carbon give widely different results in the evaporation of water or generation of steam. Theoretically, the perçentages given in the last four examples, divided by the percentage of carbon in the respective fuels, should give the relative quality of the respective steam-boilers.

Suppose the fuel used in the boilers *C* and *D* to contain 0.75 of carbon; the quality of these boilers, compared with the natural effect as a standard, will then be

$$C=\frac{48.5}{0.75}=64.6 \text{ per cent.}$$

$$D=\frac{62}{0.75}=82.6 \text{ per cent.}$$

This mode of comparing the quality of boilers with the natural effect as a standard impresses the mind at once with merits or economy of the boilers.

EVAPORATION FROM 212°.

§ 37. The comparison of steam-boiler performance with the evaporation of water from and at 212° Fahr. to steam under atmospheric pressure is a clumsy standard which repeatedly requires explanation, and even then is not always well understood. There have been many cases in which boilermakers maintained that the horse-power of their boilers should be calculated by the evaporation of water from and at 212°, while water cannot be pumped into the boiler at that temperature. When the water is heated between the feed-pump and the boiler, it is done so at the expense of the heat generated in the furnace or by the exhaust steam, and the power thus gained should not be credited to the boilermaker.

§ 38. PETROLEUM AS FUEL.

Substances.	Pounds.	Cu. ft. air.	Units of heat.
Carbon	0.84	126	12180
Hydrogen	0.16	55	9925
Petroleum	1.00	181	22105

One volume of petroleum requires 8400 volumes of air for complete combustion.

One gallon of petroleum weighs 6.7 pounds.

One pound of petroleum occupies 34.55 cubic inches.

One cubic foot of petroleum weighs 50 pounds.

Specific gravity of petroleum, 0.8.

One barrel of petroleum contains about 42 gallons, and costs in Philadelphia about six dollars, making about fifteen cents per gallon.

One barrel of petroleum weighs about 282 pounds.

Eight barrels of petroleum weigh about one ton.

One ton of petroleum costs about 45 dollars.

PERCENTAGE OF AVAILABLE HEAT OF COMBUSTION.

§ 39. When the percentage of carbon in a fuel is known (omitting hydrogen and sulphur), we can determine correctly the heat generated per pound of fuel completely consumed.

C' = fraction of carbon per pound of fuel.

The heat h generated per pound of fuel consumed will be

$h = 14500\ C'$, the gross units of heat.

h' = available heat generating steam.

$$\text{Percentage of available heat} = \frac{100\,h'}{h} \quad \ldots\ldots \quad 1$$

$$\text{Percentage of lost heat} = 100\left(1 - \frac{h'}{h}\right) \quad . \quad . \quad . \quad . \quad . \quad 2$$

The lost heat escapes with the gases of combustion through the chimney. The available heat h' is found by Formula 11, page 51.

ECONOMY OF HEATING THE FEED-WATER.

§ 40. The following Table XII. gives the percentage of gain or loss of power or fuel by different temperatures of feed-water heated or cooled. The first column contains the temperature of the feed-water at which it enters the boiler, and the top line contains the temperature from which the water is heated or cooled.

Suppose water to enter the feed-pump at 32° and to be heated to 160°, there will be 13 per cent. gained in power and fuel. When water enters the feed-pump at 60° and is heated to 150°, there will be 10 per cent. gained.

Suppose the water in the heater is 180°, which, when passing in a long pipe to the steam-boiler, is reduced to 150°, the loss will then be 3 per cent. The signs mean + for gain and − for loss:

TABLE XII.

Percentage of Power or Fuel Gained by Heating the Feed-water.

Entering boiler. Temp.	Temperature from which the Feed-water is Heated or Cooled.											
	32	40	50	60	70	80	100	120	140	160	180	200
32	0.0	−1	−1.5	−2	−3.4	−4.4	−6.5	−9	−11	−13	−16	−19
40	+1	0.0	−0.5	−1	−2.4	−3.4	−5.5	−8	−10	−12	−15	−18
50	+1.5	+0.5	0.0	−0.5	−2	−3	−4	−7	−9	−11	−14	−17
60	+2	+1	+0.5	0.0	−1.4	−2.4	−3.5	−6	−8	−10	−14	−16
70	+3.4	+2.4	+2	+1.4	0.0	−1	−3	−5	−7	−9	−13	−15
80	+4.4	+3.4	+3	+2.4	+1	0.0	−2	−4	−6	−8	−12	−14
90	+5.4	+4.4	+4	+3.4	+2	+1	−1	−3	−5	−7	−11	−13
100	+6.5	+5.5	+5	+4.4	+3	+2	0.0	−2	−4	−6	−9.5	−12
110	+7.6	+6.6	+6	+5.6	+4.2	+3.2	+1	−1	−3	−5	−8	−11
120	+8.7	+7.7	+7	+6.7	+5.3	+4.3	+2.2	0.0	−2	−4	−7	−10
130	+9.8	+8.8	+8.3	+7.8	+6.4	+5.4	+3.3	+1	−1	−3	−6	−9
140	+11	+10	+9	+9	+8	+7	+4.5	+2	0.0	−2	−5	−8
150	+12	+11	+10	+10	+9	+8	+5.5	+3	+1	−1	−3	−7
160	+13	+12	+11	+11	+10	+9	+6.5	+4	+2	0.0	−2	−6
170	+15	+14	+12	+12	+12	+11	+8	+6	+4	+2	−1	−4
180	+16	+15	+14	+14	+13	+12	+9	+7	+5	+3	0.0	−3
190	+17	+16	+15	+15	+14	+13	+10	+8	+6	+4	−1	−2
200	+19	+18	+17	+17	+16	+15	+12	+10	+8	+6	−3	0.0
212	+20	+19	+18	+18	+17	+16	+14	+11	+9	+7	−4	+1

MANAGEMENT OF FIRE IN STEAM-BOILERS.

§ 41. When the air enters under the fire-grate into the incandescent coal, its oxygen unites with the carbon and forms carbonic acid gas CO_2, which rises through the thick layer of coal and absorbs another atom of carbon, forming carbonic acid CO.

This carbonic oxide carries with it small particles of unconsumed carbon, forming smoke, which passes through the flues and tubes, and finally through the chimney into the air; the result of which is an extravagant waste of fuel.

The heat generated by forming carbonic oxide is only 30 per cent. of that generated by forming carbonic acid, together with the carrying off of unconsumed carbon in form of smoke, reduces the realized heat to a very small percentage of that due to the complete combustion of the fuel.

Therefore, in order to realize the greatest economy and effect of fuel, it must be consumed to carbonic acid, which is accomplished by keeping a very thin and even layer of fire on the grate, and by having a strong draft. For anthracite coal the thickness of the fire should be between 4 and 6 inches, and for bituminous coal from 6 to 8 inches. The carbonic acid formed will then rise to the upper surface of the fire before it can take up another atom of carbon, and the oxygen in the excess of air not utilized in the fire will unite with the unconsumed carbon rising above the coal, and form the flame.

Anthracite coal forms very little or no flame, for the reason that its hardness does not admit of faster distillation of carbon than what is immediately consumed by the oxygen of the air in contact therewith.

Bituminous coal is more easily volatilized, and the bituminous matter distills faster than it is consumed in the coal fire. The oxygen of the air, passing through the incandescent coal, consumes the gaseous carbon above the coal, forming a flame which may extend some ten feet from the furnace through the flues.

The area of entrance for air through the coal should not be less than one-fortieth ($\frac{1}{40}$) of the area of the fire-grate, and the coal layer should be of even thickness and cover completely the whole grate-surface, so that no air can enter without passing through or between incandescent coal. Should a part of the grate be uncovered with coal, a body of air will enter and reduce the temperature below that of ignition in that part of the furnace by which smoke is formed. Ashes and clinkers in the grate prevent the free access of air, and carbonic oxide and smoke are formed. An experienced fireman can see by the light in the ash-pit the condition of the fire in the grate, and he slices the fire accordingly. When the furnace is charged, the coal should be spread evenly all over the fire, and the furnace doors should not be kept open longer than is necessary for the charge.

PRODUCTS OF COMBUSTION.

§ 42. The term "products of combustion" should mean only the binary compound of oxygen and combustibles formed in the operation of combustion, such as carbonic oxide, carbonic acid, steam and sulphurous acid; but, practically, all the gases in the furnace, including nitrogen and smoke, are termed products of combustion. When hydrogen is consumed in the furnace and forms steam, that steam is then a product of combustion; but when evaporated from moisture in the fuel, it is not a product of combustion in the furnace.

TABLE XIII.

Properties of Products of Combustion.

Gases of Combustion.	Atomic		Specific		Weight and volume at 60°.	
	Symbol.	Weight.	Gravity.	Volume.	lbs. per cu. ft.	cu. ft. per lb.
Atmospheric air........	N_2O	36	1.	1.	0.0760	13.158
Oxygen....................	O	8	1.104	0.9058	0.0839	11.9189
Nitrogen..................	N	14	0.972	1.0288	0.0740	13.5135
Hydrogen	H	1	0.0693	14.430	0.000267	189.86
Carbon	C	6	0.8380	1.1933	0.06369	15.701
Sulphur	S	16	1.123	0.8904	0.0853	11.723
Carbonic oxide..........	CO	14	0.972	1.0288	0.0740	13.5135
Carbonic acid............	CO_2	22	1.527	0.6549	0.11505	8.6900
Steam	HO	9	0.625	1.6	0.0475	21.0526
Carburetted hydrogen...	H_2C	8	0.555	1.8018	0.04218	23.7079
Bicarburetted hydrogen	H_2C_2	14	0.98	1.0204	0.07448	13.4264
Nitrous oxide............	NO	22	1.525	0.6557	0.1159	8.6281
Sulphurous acid	SO_2	32	2.247	0.4450	0.19077	5.2415

GRATE-BARS.

§ 43. The proportion of thickness of grate-bars to the air-space between them varies between 1 and 3 to 1, depending on the kind of fuel used on the grate—that is to say, the area of air-passage between the bars varies between 25 and 50 per cent. of the grate-surface.

The following table gives the spaces between the grate-bars in fractions of an inch, as generally used for different kinds of fuel.

SPACE BETWEEN GRATE-BARS.

Lehigh anthracite pea coal	$\frac{1}{4}$	of an inch.
Schuylkill " " "	$\frac{3}{8}$	" "
Lehigh " chestnut coal	$\frac{3}{8}$	" "
Lehigh " stove "	$\frac{1}{2}$	" "
Lehigh " broken "	$\frac{5}{8}$	" "
Cumberland bituminous coal.	$\frac{3}{4}$	" "
Ordinary wood	$\frac{3}{4}$ to 1	"
Sawdust	$\frac{3}{16}$ to $\frac{1}{4}$	"

SMOKE-BURNING.

§ 44. The burning of smoke has, since the time of Watt, received a great deal of attention, but not with much success, owing, first, to insufficient knowledge of the chemistry of smoke, which in Watt's time was not sufficiently developed for that purpose; and secondly, the physical properties of smoke have not been properly considered in the attempt to burn smoke.

When the science of chemistry was sufficiently advanced to enable us to determine correctly the elements of combustion and of smoke, we have still not fully considered the physical properties bearing upon the success in smoke-burning.

It is well known that smoke consists of small particles of carbon mixed with carbonic oxide, both of which are combustibles, with a sufficient supply of oxygen at a temperature above that of ignition between 700° and 800° Fahr. It appears, therefore, that a sufficient supply of air among the smoke in the furnace would accomplish the object, but unfortunately such has not been the result.

Suppose a case of one pound of carbon being consumed by the oxygen of 103 cubic feet of air, which, according to Table X., will form

Carbonic oxide	CO	= 1.5166 lbs. =	20.494	cubic feet.
Carbonic acid	CO_2	= 1.2833 lbs. =	11.126	" "
Nitrogen	N	= 6.2075 lbs. =	81.370	" "
Products of combustion		= 9.0074 lbs. =	112.990	cubic feet.

The volume is here taken at 60° Fahr.; but at a temperature above that of ignition, say 800°, the volume of the products of combustion will be $2.5 \times 113 = 282.5$ cubic feet. (See Law of Gases.)

Of this volume only $2.5 \times 20.5 = 51.25$ cubic feet is combustible or carbonic oxide, which requires the oxygen of $76.5 \times 1.5166 = 115.9$ cubic feet of air at 60° for combustion to carbonic acid.

The gases of combustion are not chemically combined, but mechanically mixed in the furnace, and arrange themselves into layers according to their specific gravity, the lightest occupying the top and the heaviest the bottom of the furnace or flues. The specific gravity of nitrogen and carbonic oxide being alike, these two gases will mix; but the nitrogen, which is a non-supporter of combustion, occupies four times the volume of that of the combustible carbonic oxide.

We see here the difficulty of uniting the oxygen of 116 cubic feet of air at 60° with $2.5 \times 11.126 = 37.8$ cubic feet of carbonic oxide, which is already mixed with $2.5 \times 81.37 = 203.42$ cubic feet of nitrogen; therefore the burning of carbonic oxide to carbonic acid by additional supply of air to the furnace may be considered very difficult, if not impossible.

When the carbonic oxide is mixed with free carbon at a temperature above that of ignition, the oxygen of a supply of air is easier united with these combustibles, but even then the large quantity of nitrogen will interfere with that combustion.

The smoke is formed first when the temperature of the products of combustion is reduced below that of ignition, before which time the free carbon is incandescent.

In most of the attempts made to burn smoke by additional supply of air, the air has been admitted under the gases of combustion—that is, from behind or from the top of the bridge, where it first comes in contact with the carbonic acid, and perhaps sulphuric acid, which prevents the air from being mixed with the combustibles before the temperature is reduced below that of ignition.

The admission of a small quantity of air through the fire-door or to the upper part of the furnace has proven partly successful in burning some smoke, but the most economical combustion of the fuel is when the furnace and fire are so arranged that the fuel is completely consumed by the air entering through the grate into the fire.

NATURAL FURNACE-DRAFT.

§ 45. The natural draft to a furnace is caused by the column of heated gases in the chimney being lighter than an equal column of the surrounding air. The weight of a cubic foot of dry air at 60° is 532 grains; and suppose the hot gases in the chimney to weigh 286 grains per cubic foot, then a chimney of one square foot section, and say 50 feet high, would contain 50 cubic feet, and the weight of the hot gases

$50 \times 286 = 14300$ grains. The weight of an equal column of air at 60° would weigh $50 \times 532 = 26600$ grains, and $26600 - 14300 = 12300$ grains, which will be the pressure per square foot of the draft.

The height of a column of air answering to this pressure is $12300 : 532 = 23.12$ feet. The velocity of the draft through the fire (which is the smallest aperture of entrance) will be equal to that a body would attain by falling vertically a height of 23.12 feet—namely, 36.44 feet per second.

The combustion of one pound of carbon produces by 153 cubic feet of air,

Carbonic acid	$CO_2 = 3.6666$ lbs. =	31.86	cubic feet.
Nitrogen	$N = 8.9455$ lbs. =	120.87	" "
Total	$= 12.6121$ lbs. =	152.73	" "

We see here that the volume of the gases of combustion is nearly equal to that of the air supplied, but their specific gravity is slightly more—namely, as $12.612 : 11.552 = 1.0918$.

Some carbonic oxide, which is lighter than air, always accompanies the gases, for which we may with safety assume the sp. gr. of the gases of combustion to be equal to that of air of the same temperature. Therefore the sp. gr. of the hot gases in the chimney will be equal to the reciprocal of the volume expansion by heat, which is denoted by x in the Table XXX. for law of gases.

For a temperature of 500° of the gases in the chimney the volume is $x = 1.9491$, which reciprocal is 0.51308, the required specific gravity of the gases.

The height of the chimney is to the height of a column of cool air of equal weight to that of the hot air as $1 : \left(1 - \frac{1}{x}\right)$.

A = section area of the chimney, and

⊟ = area of the fire-grate in square feet.

V = velocity of the air through the fire.

v = ascending velocity of the gases in the chimney.

H = height of the chimney in feet above the fire-grate.

The area for passing the air through the fire should be about one-fortieth ($\frac{1}{40}$) of the area of the fire-grate.

The area of the chimney is generally made about 0.16 of the area of the fire-grate.

$$V = 5\sqrt{H\left(1 - \frac{1}{x}\right)} \quad . \quad . \quad . \quad . \quad . \quad 1$$

The theoretical coefficient should be 8 instead of 5.

$$v = \frac{\boxminus V}{40\,A} = \frac{\boxminus}{8\,A}\sqrt{H\left(1 - \frac{1}{x}\right)} \quad . \quad . \quad . \quad 2$$

Example 1. The height of a chimney is $H = 75$ feet, and the temperature of the ascending gases 450°. Required the velocity of the air through the fire?

Formula 1. $V = 5\sqrt{75\left(1 - \frac{1}{1.8477}\right)} = 29.33$ feet per second.

Example 2. Required the velocity of the ascending gases in the chimney when $\boxminus = 36$ square feet and $A = 5.76$ square feet?

Formula 2. $v = \frac{36}{8 \times 5.76}\sqrt{75 + 0.4588} = 4.58$ feet per second.

It is assumed in these examples that the temperature of the air is 32°, but for other temperatures of the air a corresponding reduction should be made of the temperature of the hot gases; for example, when the air is 75° and the gases 450°, then $75 - 32 = 43°$, and $450 - 43 = 407°$, the temperature for the velocity of the ascending gases.

The factor $\left(1 - \frac{1}{x}\right)$ in the Formulas 1 and 2 denoted by z in Table XXX. is

$$z = \left(1 - \frac{1}{x}\right) = \frac{\mathfrak{T} - \mathfrak{t}}{493 + \mathfrak{T} - \mathfrak{t}}, \text{ in which,} \quad . \quad . \quad 3$$

$\mathfrak{T}$ = temperature of the ascending gases in the chimney, and
$\mathfrak{t}$ = temperature of the surrounding air.

$$V = 5\sqrt{Hz} \quad . \quad . \quad . \quad . \quad 4$$

$$v = \frac{\boxminus}{8\,A}\sqrt{Hz} \quad . \quad . \quad . \quad . \quad 5$$

As in Formula 1, the coefficient 5 in Formula 4 should be 8 by the acceleration of gravity $V = 8\sqrt{gS}$; but the friction and turning of the gases amongst the incandescent fuel and returning tubes reduce the velocity over 30 per cent., for which reason the coefficient is reduced from 8 to 5.

WATER-GAUGE FOR CHIMNEY DRAFT.

§ 46. The difference of pressure between the hot gases in the chimney and the surrounding atmosphere is very small, and is therefore measured by a column of water.

A cubic foot of water at 32° Fahr. weighs 62.387 pounds, whilst a cubic foot of air of the same temperature weighs only 0.0804186 of a pound; therefore a column of air must be 62.387 : 0.0804186 = 766.25 times higher than a column of water for the same pressure.

The height of a column of air corresponding to the difference of pressure in and outside the chimney is

$$H' = H\left(1 - \frac{1}{x}\right) \quad . \quad . \quad . \quad . \quad . \quad 1$$

x = volume expansion of gases by heat corresponding to the temperature of the gases in the chimney from the Table XXX. of law of gases.

H = height of the chimney in feet above grate.

This height H', divided by 766.25, gives the height of a column of water of equal pressure, and multiplied by 12 gives the height in inches, denoted by I.

$$I = \frac{12H}{766.25}\left(1 - \frac{1}{x}\right) = \frac{H}{63.854}\left(1 - \frac{1}{x}\right) \quad . \quad . \quad 2$$

$$I = \frac{H(T - t)}{63.854(493 + T - t)} \quad . \quad . \quad . \quad 3$$

The following Table XIV. is calculated from this formula for different temperatures T of the gases in the chimney, and that of the air $t = 32$, and for different heights H of chimney.

The water-gauge should be placed as near the level of the fire-grate as practicable.

TABLE XIV.

Water-gauge in Inches for Chimney-draft.

Height of Chimney.	Temperatures T of Gases in the Chimney.						
	400	450	500	550	600	700	800
H.	I.	I.	I.	I.	I.	I.	I.
10	0.0669	0.0718	0.0762	0.0802	0.0838	0.0901	0.0974
15	0.1000	0.1077	0.1143	0.1203	0.1257	0.1356	0.1430
20	0.1338	0.1437	0.1525	0.1604	0.1677	0.1802	0.1907
30	0.2008	0.2155	0.2287	0.2407	0.2515	0.2703	0.2861
40	0.2678	0.2874	0.3050	0.3209	0.3354	0.3604	0.3815
50	0.3346	0.3592	0.3812	0.4011	0.4192	0.4505	0.4768
60	0.4016	0.4311	0.4575	0.4814	0.5031	0.5406	0.5722
70	0.4685	0.5029	0.5337	0.5616	0.5870	0.6307	0.6676
80	0.5354	0.5748	0.6100	0.6418	0.6709	0.7208	0.7630
90	0.6024	0.6466	0.6862	0.7221	0.7547	0.8109	0.8584
100	0.6693	0.7185	0.7625	0.8023	0.8385	0.9010	0.9537
125	0.8366	0.8981	0.9531	1.0028	1.0481	1.1262	1.1921
150	1.0039	1.0777	1.1437	1.2034	1.2577	1.3515	1.4305
175	1.1712	1.2573	1.3343	1.4039	1.4673	1.5767	1.6689
200	1.3386	1.4370	1.5250	1.6046	1.6770	1.8020	1.9074
250	1.6732	1.7962	1.9062	2.0057	2.0962	2.2525	2.3842
300	2.0079	2.1555	2.2875	2.4069	2.5155	2.7030	2.8611
400	2.6772	2.8740	3.0500	3.2092	3.3540	3.6040	3.8148

§ 47. QUANTITY OF AIR BY NATURAL DRAFT.

Q = cubic feet of air passing through the fire per hour by natural draft.

$$Q = 90 \boxminus V = 450 \boxminus \sqrt{H\left(1-\frac{1}{x}\right)}. \quad . \quad . \quad 1$$

The average quality of coal may be assumed to contain 0.75 of pure carbon, and $153 \times 0.75 = 115$ cubic feet of air required per pound of coal consumed. For safety say 140 cubic feet.

L = pounds of coal consumed per hour per square foot of grate.

$$Q = 140 L \boxminus. \quad . \quad . \quad . \quad . \quad . \quad . \quad 2$$

$$L = 3.2\sqrt{H\left(1-\frac{1}{x}\right)}. \quad . \quad . \quad . \quad 3$$

Example 3. How much coal will be consumed per hour per square foot of grate by a chimney of $H = 60$ feet high, the temperature of the ascending gases being 500° ?

Formula 3. $L = 3.2\sqrt{60\left(1 - \frac{1}{1.949}\right)} = 17.28$ pounds.

The height of the chimney required for the combustion of L pounds of coal per hour per square foot of grate will be

$$H = \frac{L^2}{10.29\left(1 - \frac{1}{x}\right)} \quad \ldots \quad 4$$

LOSS OF HEAT BY THE ESCAPING GASES OF COMBUSTION.

§ 48. The heat carried off by the gases of combustion is lost for the generation of steam, but utilized for creating draft to the furnace. The higher the chimney is, the more will that heat be utilized for creating draft. The economy consists in making the chimney high and reducing the temperature of the ascending gases by absorbing more of the heat for evaporation.

The specified heat of the gases of combustion averages 0.25. See Specific Heat. The weight of the gases per pound of carbon consumed to carbonic acid is 12.612 pounds, and the heat carried off will be

$$h = 12.612 \times 0.25\,(T - t) \quad \ldots \quad 1$$

$$h = 3.153\,(T - t) \quad \ldots \quad 2$$

c = fraction of carbon per pound of coal.
L = pounds of coal consumed per hour per square foot of grate.
h = units of heat passing through the chimney per hour.

$$h = 3.153cL \equiv (T - t) \quad \ldots \quad 3$$

The percentage of heat lost by the escaping gases will then be

$$0.02175\,(T - t) \quad \ldots \quad 4$$

Example 4. The temperature of the ascending gases being $T = 480°$, and that of the surrounding air $t = 72°$, required the percentage of heat lost through the chimney?

$$0.02175(480 - 72) = 8.87 \text{ per cent.}$$

It is supposed in this example that all the carbon is perfectly consumed to carbonic acid.

TEMPERATURE OF THE GASES IN THE CHIMNEY.

§ 49. This is a very difficult problem to solve theoretically, on account of the various circumstances involved therein making a very complicated mathematical demonstration, the result of which would probably not give a closer result than does the following formula, which is set up from practice; namely,

$$T = 300\sqrt{\frac{\boxminus\sqrt{(p+2)(H+2)}}{\boxminus + \square}}. \quad . \quad . \quad 1$$

T = temperature of the gases when entering the chimney.

Example 1. A steam-boiler of $\boxminus = 96$ square feet fire-grate and $\square = 2880$ square feet of heating surface is connected with a chimney $H = 47$ feet high. Steam pressure $p = 62$ pounds to the square inch. Required the temperature of the gases in the chimney?

$$T = 300\sqrt{\frac{96\sqrt{(62+2)(47+2)}}{96+2880}} = 403.2° \text{ Fahr.}$$

By this formula we can find the temperature in any part of the flues or tubes by subtracting that part of the heating surface which the fire has not reached, or by taking the heating surface exposed to the fire up to the point where the temperature is required.

Example. Required the temperature at the bridge in the boiler of the preceding example, in which the heating surface in the furnaces alone is $\square = 245$ square feet?

$$T = 300\sqrt{\frac{96\sqrt{(62+2)(47+2)}}{96+245}} = 1191° \text{ Fahr.}$$

It is assumed in this formula and examples that the cross-section of the chimney is $A = 0.16\boxminus$.

The temperature in the chimney ought not to be more than 100° above that of the steam in the boiler, and the heating surface not more than $\square = 36\boxminus$.

The proper proportion between the fire-grate and heating surface depends upon the steam-pressure, or rather the temperature of the steam and that of the gases in the chimney. When the temperature of the latter is reduced below that of the former, heat is conducted from the water back into the flue, which operation is a waste of fuel, material and labor in the first construction of such boilers.

In locomotive boilers with very long and narrow tubes and exhaust draft in the chimney, the temperature of the gases has often been reduced below that of the water and steam in the boiler, the result of which is a waste of fuel.

In marine boilers the heating surface rarely exceeds 36 ⊟, and the temperature of the gases in the chimney is then about 100° over that of the steam in the boiler.

Stationary boilers are sometimes made with heating surface = 50 ⊟, and the temperature of the gases in the chimney has been reduced below that of the steam; but the water evaporated per pound of combustibles has been less than with smaller proportions of heating surface.

For very low steam-pressure the heating surface may advantageously be made = 50 ⊟.

When there is no heating surface, but the chimney is connected directly to the fire-grate, so that all the heat ascends in the chimney, the temperature will then be

$$T = 300\sqrt{7\sqrt{H+2}}. \quad . \quad . \quad . \quad 2$$

Example 2. Required the temperature in a chimney $H = 62$ feet high, connected directly with the fire-grate without water-heating surface, but that all the heat passes up the chimney?

$$T = 300\sqrt{7\sqrt{62+2}} = 2244.5 \text{ Fahr.}$$

§ 50. TEMPERING OF STEEL.

The temperature of the gases in the chimney depends much upon the construction of the boiler and the proportion of fire-grate and heating surface. A simple way of measuring this temperature approximately is by inserting a polished iron wire about ¼ of an inch in diameter; the color of tempering will show the temperature, corresponding with the following table.

The property of heat to color steel or iron can be applied for ascertaining the temperature in flues and chimneys of steam-boilers, and for other temperatures limited between 430° and 600° Fahr.

Yellow, very faint, for lancets	430°
" pale straw, for razors, scalpels	450°
" full, for penknives and chisels for cast iron	470°
Brown, for scissors and chisels for wrought iron	490°
Red, for carpenters' tools in general	510°
Purple, for fine watch-springs and table-knives	530°
Blue, bright, for swords, lock-springs	550°
" full, for daggers, fine saws, needles	560°
" dark, for common saws	600°

EVAPORATION OF POUNDS OF WATER PER HOUR PER SQUARE FOOT OF HEATING SURFACE.

§ 51. The evaporation per heating surface varies directly as the $1\frac{1}{2}$ power of the difference between the temperature of the gases of combustion and that of the water in the boiler. The temperature of the gases is determined by Formula 1, paragraph 49, and the temperature of the water is the same as that corresponding to the steam-pressure. The evaporation per heating surface will therefore be different in different parts of the boiler.

h = units of heat passed through each square foot of heating surface per hour.

H = units of heat per pound of steam generated. (See Steam Table, Nystrom's *Pocket-Book.*)

T = temperature of the gases of combustion at the place in the boiler where the rate of evaporation is calculated.

t = temperature of the water or steam.

lbs = pounds of water evaporated per hour per square foot of heating surface at the place where the temperature of the gases is T.

$$\text{Units of heat,} \quad h = 0.505\sqrt{(T - t)^3}. \quad . \quad 1$$

$$\text{Evaporation,} \quad \text{lbs.} = \frac{0.505\sqrt{(T - t)^3}}{H}. \quad . \quad 2$$

Example 1. The temperature in a boiler furnace is $T = 1200°$, and steam pressure 80 pounds to the square inch, which corresponds to $t = 324$ temperature of the steam. Required the units of heat passing through each square foot of heating surface per hour? and how much water will be evaporated per square foot of heating surface per hour?

$$\text{Units of heat,} \quad h = 0.505\sqrt{(1200 - 324)^3} = 13093.$$

$$\text{Evaporation,} \quad \text{lbs.} = \frac{13093}{1180.7} = 11.09 \text{ pounds.}$$

Example 2. In the same steam-boiler as in the preceding example, the temperature of the gases entering the chimney is $T = 460°$. Required the evaporation per square foot of heating surface at the end of the boiler where the gases of combustion enter the chimney?

$$\text{Evaporation,} \quad \text{lbs} = \frac{0.505\sqrt{(460° - 324)^3}}{1180.7} = 0.678 \text{ of a pound.}$$

The rate of evaporation can thus be calculated in any part of the boiler by first calculating the temperature T from Formula 1, in paragraph 49.

FRESH WATER CONDENSERS.

§ 52. Fresh water condensers are generally made of brass tubes about $\frac{5}{8}$ of an inch diameter and tinned outside.

h = units of heat conducted per hour through each square foot of tubes.

T = temperature of the steam entering the condenser.

t = temperature of the water entering the condenser.

⊟ = area of fire-grate in square feet.

□ = heating surface in square feet.

A = tubular area in square feet in the condenser, required to condense the steam generated by the boiler ⊟ □.

$$h = 0.6\sqrt{(T - t)^3}. \quad . \quad . \quad . \quad . \qquad 1$$

$$A = 3.5\sqrt{⊟ □}. \quad . \quad . \quad . \quad . \qquad 2$$

Example 2. How much tubular condensing surface is required for a boiler of ⊟ = 128 square feet fire-grate, and heating surface □ = 3850 square feet?

Condensing surface, $A = 3.5\sqrt{128 \times 3850} = 2457$ square feet.

SAFETY-VALVES.

§ 53. The area of a safety-valve should be sufficiently large to let out all the steam the boiler can generate without increasing the normal working pressure of the boiler, and without the valve lifting more than one-forty-eighth ($\frac{1}{48}$) of its diameter.

A = area in square inches of the inner circle of the valvesit.

a = area through which the steam escapes, which is equal to the circumference of the inner circle of the valvesit multiplied by the height the valve is lifted.

p = steam-pressure in pounds per square inch above that of the atmosphere.

$\mathfrak{P}$ = weight in a fraction of a pound per cubic foot of the steam of pressure p. (See Steam Table, Nystrom's *Pocket-Book.*)

$\mathfrak{V}$ = steam-volume compared with that of its water at 32° Fahr.

H = height in feet of a column of steam of one square foot section, which weight would be equal to the steam-pressure per square foot, or $144\,p$.

V = velocity in feet per second of the steam through the safety-valve.

The weight of a column of steam of height H and weight per cubic foot $\mathfrak{P}$ will then be $H\,\mathfrak{P}$.

That is to say, $$144\,p = H\,\mathfrak{P}. \qquad 1$$

Height of column, $$H = \frac{144\,p}{\mathfrak{P}}. \qquad 2$$

Velocity of steam, $$V = 8\sqrt{H} = 96\sqrt{\frac{p}{\mathfrak{P}}} \qquad 3$$

Q = cubic feet of steam discharged through the safety-valve per second.

$$Q = \frac{a\,V}{144} = \frac{8\,a}{144}\sqrt{H} = \frac{96\,a}{144}\sqrt{\frac{p}{\mathfrak{P}}} = \frac{2}{3}a\sqrt{\frac{p}{\mathfrak{P}}} \qquad 4$$

That is to say, the steaming capacity of the boiler in cubic feet of steam per second should not exceed

$$Q = \tfrac{2}{3}a\sqrt{\frac{p}{\mathfrak{P}}} \qquad 5$$

The steaming capacity of a boiler fired with a given kind or quality of fuel depends upon the area of the fire-grate and heating surface. With natural draft the average evaporation of water of 32° to Q cubic feet of steam per second in ordinary boilers is

$$Q = \frac{\forall\sqrt{\boxminus\,\square}}{9000} \qquad 6$$

This should be equal to the escape of steam through the safety-valve, Formula 5, or

$$Q = \frac{2}{3}a\sqrt{\frac{p}{\mathfrak{P}}} = \forall\frac{\sqrt{\forall\,\boxminus\,\square}}{9000} \qquad 7$$

From this formula we obtain the requisite area a of the safety-valve for letting out all the steam the boiler can generate—namely,

$$a = \frac{3\,\forall\sqrt{\boxminus\,\square}}{2\times 9000}\sqrt{\frac{\mathfrak{P}}{p}} = \frac{\forall\sqrt{\boxminus\,\square}}{6000}\sqrt{\frac{\mathfrak{P}}{p}} \qquad 8$$

Allowing for the contraction of the steam through the valve,	35 per cent.
For guiding wings of the valve	20 " "
For steam generation, Formula 6	20 " "
Reduction for safety	75 " "

Limiting the valve to lift only one-forty-eighth of its diameter, the coefficient 6000 in Formula 8 will be reduced to 288, when A is the area of the inner circle of the valvesit.

$$A = \frac{\mathfrak{V}\sqrt{\boxminus\square}}{288}\sqrt{\frac{p}{\mathfrak{W}}} \quad . \quad . \quad . \quad . \quad 9$$

This should be the reliable formula for the requisite area of the safety-valve of a steam-boiler.

Example 9.—A steam boiler of $\boxminus = 130$ square feet of fire-grate and $\square = 3372$ square feet of heating surface, carrying $p = 49$ pounds of steam-pressure per square inch above that of the atmosphere. Required the area A of the safety-valve?

The steam volume at $p = 49$ is $\mathfrak{V} = 403.29$, and weight per cubic foot of steam $\mathfrak{W} = 0.15469$ of a pound. The area of the safety-valve will then be

$$A = \frac{403.29\sqrt{130 \times 3372}}{288}\sqrt{\frac{0.15469}{49}} = 52.092 \text{ square inches.}$$

The Formula 9 can be reduced to a very simple form by the aid of a table, for which make

$$M = 288\sqrt{\frac{p}{\mathfrak{W}}} \quad . \quad . \quad . \quad . \quad 10$$

$$N = \sqrt{\frac{p}{\mathfrak{W}}} \quad . \quad . \quad . \quad . \quad 11$$

SIT OF SAFETY-VALVES.

§ 54. The sit of a safety-valve should be flat, and not conical. A flat joint is easier ground and kept tight than a conical one. The width of a valvesit should not be more than one-tenth ($\frac{1}{10}$) of the cube root of the diameter of the valve, and even one-sixteenth will answer the purpose.

For conical valves the area should be

$$A = \frac{\sqrt{\boxminus\square}}{M \cos.v.} \quad . \quad . \quad . \quad . \quad 1$$

v = angle of the valvesit to the plane of the valve.

For an angle of 45° $\cos.45° = 0.707$, and,

$$A = \frac{\sqrt{\boxminus\square}}{0.707\,M} \quad . \quad . \quad . \quad . \quad 2$$

The columns M and N, in the following Table XV., are calculated from the Formulas 10 and 11 for different steam-pressures in the first column p.

The formula for area of safety-valves will then be simply

$$A=\frac{\sqrt{\boxminus\square}}{M.} \quad . \quad . \quad . \quad . \quad 3$$

Example 3.—Required the area of a safety-valve for a boiler of $\boxminus=36$ square feet fire-grate, and $\square=1024$ square feet heating surface, to carry $p=85$ pounds steam-pressure? (See Table XV.)

$$A=\frac{\sqrt{36\times1024}}{20.52}=9.375 \text{ square inches.}$$

If the same boiler should be limited to $p=20$ pounds steam-pressure, the area of the safety-valve should be,

$$A=\frac{\sqrt{36\times1024}}{6.107}=31.44 \text{ square inches.}$$

The steam-volume in the following table is calculated from Fairbairn's formula.

VELOCITY OF STEAM FORCED BY ITS PRESSURE INTO AIR OR VACUUM.

§ 55. The velocity of steam forced by its pressure into the atmosphere is

$$V=96\sqrt{\frac{p}{\mathfrak{P}}} \quad . \quad . \quad . \quad . \quad 1$$

When the steam passes into a vacuum, the velocity will be

$$V=96\sqrt{\frac{P}{\mathfrak{P}}} \quad . \quad . \quad . \quad . \quad 2$$

$\mathfrak{P}$ = weight in pounds per cubic foot of steam.

P = pressure per square inch above vacuum.

When the steam passes into a partial vacuum of pressure p'—that is, the difference between the atmospheric pressure and that into which the steam passes—the velocity will be

$$V=96\sqrt{\frac{p+p'}{\mathfrak{P}}} \quad . \quad . \quad . \quad 3$$

TABLE XV.

Area of Safety-valves and Velocity of Steam Passing into the Air.

Steam pressure.	$\frac{288}{\mathfrak{V}}\sqrt{\frac{p}{\mathfrak{P}}}=$		$\sqrt{\frac{p}{\mathfrak{P}}}=$	Velocity. $96\,N=$	Fairbairn's Steam volume.	Weight per cubic foot of steam.
p.	M.	*Logarithms.*	N.	V.	$\mathfrak{V}$.	$\mathfrak{P}$
5	2.333	0.3680283	9.883	948.77	1219.7	0.05119
10	3.675	0.5652855	12.56	1205.7	984.23	0.06338
15	4.911	0.6911552	14.09	1352.6	826.32	0.07550
20	6.107	0.7858060	15.12	1451.5	713.08	0.08749
25	7.274	0.8617983	15.86	1522.5	627.91	0.09936
30	8.427	0.9256803	16.43	1577.3	561.50	0.11111
35	9.570	0.9089105	16.89	1621.4	508.29	0.12273
40	10.70	1.0292700	17.26	1656.9	464.69	0.13128
45	11.82	1.0726430	17.58	1707.6	428.42	0.14566
50	13.21	1.1208622	17.85	1713.6	397.51	0.15694
55	14.04	1.1473753	18.09	1734.8	371.07	0.16812
60	15.14	1.1800772	18.30	1756.8	348.15	0.17919
65	16.23	1.2103496	18.49	1774.0	328.06	0.19015
70	17.32	1.2385479	18.66	1791.3	310.36	0.20101
75	18.39	1.2647646	18.82	1806.7	294.61	0.21185
80	19.46	1.2891428	18.97	1821.1	280.50	0.22241
85	20.52	1.3121774	19.10	1833.6	267.80	0.23296
90	21.59	1.3342526	19.23	1846.1	256.31	0.24340
95	22.66	1.3552599	19.35	1857.6	245.86	0.25375
100	23.73	1.3752764	19.47	1869.1	236.31	0.26400
105	24.78	1.3941013	19.57	1878.7	227.56	0.27421
110	25.81	1.4117624	19.67	1888.3	219.50	0.28422
115	26.85	1.4289443	19.76	1897.0	212.07	0.29419
120	27.88	1.4452367	19.86	1906.6	205.18	0.30406
125	28.91	1.4610481	19.95	1915.2	198.78	0.31385
130	29.95	1.4763323	20.05	1924.8	192.83	0.32354
135	30.99	1.4912226	20.14	1933.4	187.26	0.33315
140	32.11	1.5066060	20.24	1943.0	181.69	0.34276

a = area in square inches through which the steam escapes.
Q = cubic feet of steam passing through the opening a per second.
m = coefficient of contraction of the steam-jet, which varies from 0.64 to 1. For steam escaping through valves or cocks the coefficient can be taken to $m = 0.75$.

$$Q = \frac{m\,a\,V}{144} \quad . \quad . \quad . \quad . \quad . \quad . \quad 4$$

Placing $m = 0.75$, we have

$$Q = 0.5\,a\sqrt{\frac{p}{\mathfrak{P}}} \quad . \quad . \quad . \quad . \quad . \quad . \quad 5$$

$$Q = 0.5\,a\sqrt{\frac{P}{\mathfrak{P}}} \quad . \quad . \quad . \quad . \quad . \quad . \quad 6$$

$$Q = 0.5\,a\sqrt{\frac{p+p'}{\mathfrak{P}}} \quad . \quad . \quad . \quad . \quad . \quad . \quad 7$$

§ 56. When steam passes into air of atmospheric pressure, the velocity and cubic feet of steam discharged per second are easily calculated by the aid of Table XV.—namely,

Velocity, $V = 96\,N$ 8
Cubic volume, $Q = 0.5\,a\,N$ 9

Example 8. Required the velocity of steam passing from a boiler and under $p = 65$ pounds pressure?

$$V = 96 \times 18.49 = 1775.04 \text{ feet per second.}$$

Example 9. Required the volume of that steam passing through an orifice of $a = 1.5$ square inches?

$$Q = 0.5 \times 1.5 \times 18.49 = 13.867 \text{ cubic feet per second.}$$

Example 1. Required the velocity V of steam of pressure $p = 65$ pounds to the square inch above that of the atmosphere, issuing from the boiler into the air? and how many cubic feet will be discharged per second through an opening $a = 0.75$ of a square inch? When the opening is through a thin plate in which the steam-jet is contracted on all sides, the coefficient is $m = 0.64$.

Velocity, $$V = 96\sqrt{\frac{65}{0.19015}} = 1775 \text{ feet per second.}$$

Steam discharged, $$Q = \frac{0.64 \times 0.75 \times 1775}{144} = 5.91 \text{ cubic feet per second.}$$

Example 6. What quantity of steam of pressure $P=85$ pounds to the square inch above vacuum will pass through a cock of $a=0.45$ of a square inch into a vacuum?

$$Q=0.5\times0.45\sqrt{\frac{85}{0.2010}}=4.627 \text{ cubic feet.}$$

Example 8. Steam of pressure $p=45$ pounds to the square inch above the atmosphere is passing into a partial vacuum of 18.33 inches mercury, or $p'=9$ pounds to the square inch. Required the velocity of the steam? and how much will pass through the opening of $a=1.25$ square inches, the coefficient of contraction being $m=0.8$?

$$V=96\sqrt{\frac{45+9}{0.14566}}=1852.7 \text{ feet per second.}$$

$$Q=\frac{0.8\times1.25\times1852.7}{144}=11.68 \text{ cubic feet.}$$

The horse-power per volume of steam consumed per hour is given by Formula 1, § 23, in which

$$Q=W(\dot{V}-1) \quad . \quad . \quad . \quad . \quad 10$$

$$\text{HP}=\frac{3600\,p\,Q}{13748.4}=\frac{p\,Q}{3.819} \quad . \quad . \quad . \quad 11$$

$$Q=\frac{3.819\,\text{HP}}{p} \quad . \quad . \quad . \quad . \quad . \quad 12$$

$$Q=\frac{m\,a\,V}{144}=\frac{96\,m\,a}{144}\sqrt{\frac{p}{\mathfrak{P}}}=\frac{2}{3}m\,a\sqrt{\frac{p}{\mathfrak{P}}}=\frac{3.819\,\text{HP}}{p}$$

$$\text{HP}=\frac{m\,a\,p}{5.7285}\sqrt{\frac{p}{\mathfrak{P}}} \quad . \quad . \quad . \quad . \quad 13$$

This formula gives the horse-power of steam of pressure p escaping from a boiler through an opening a.

Example 13. What horse-power is required to blow a steam-whistle 4 inches in diameter, when the opening is 0.005 of an inch, the steam-pressure being $p=60$ pounds to the square inch above atmospheric pressure?

The area of the steam-whistle is

$$a=4\times3.14\times0.005=0.0628 \text{ of a square inch.}$$

In this case the steam passes through a taper opening, for which the coefficient $m=1$.

$$HP=\frac{0.0628\times 60}{5.7285}\sqrt{\frac{60}{0.17919}}=12 \text{ horse-power.}$$

This seems to be a very high horse-power for a steam-whistle, but it is nevertheless true under the conditions assumed.

§ 57. HORSE-POWER OF STEAM-ENGINES BY VOLUME OF STEAM.

C = cubic feet of full steam used in each single stroke in the steam-cylinder.
n = double strokes of piston per minute.
p = steam-pressure in pounds per square inch.

$$HP=\frac{2\,n\,C\,p}{3.819\times 60}=\frac{n\,C\,p}{114.57}. \qquad 1$$

Example 1. The cubic capacity of a steam cylinder is $C=6.5$ cubic feet, and the piston makes $n=45$ double strokes per minute with a steam-pressure of $p=70$ pounds to the square inch. Required the horse-power of the engine?

$$HP=\frac{45\times 6.5\times 70}{114.57}=178.7 \text{ horse-power.}$$

This is the horse-power of the high-pressure engine working with full steam.

If the horse-power of the same engine is calculated in the ordinary way, it will be 180.7, or one horse-power more than in the example, which is the power consumed by the force-pump feeding the boiler with water.

When the steam is expanded in the cylinder, C means the volume of the full steam, and the horse-power of the full steam multiplied by $1+hyp.log.$ of the expansion is the horse-power of the expanded steam.

STEAM-PRESSURE AND REVOLUTIONS.

§ 58. When the dimensions of the boiler and engines are given, to find the relation between steam-pressure and revolutions of the engine.

V = steam-volume compared with that of its water at 32° for the given steam-pressure.

$\mathfrak{V}$ = cubic feet of unexpanded steam used in each revolution of the engine or engines.

n = number of revolutions per minute of the engine.

R = correction for temperature of feed-water, Table V.

r = correction for height of chimney, Table VII.

$$\mathcal{V} = \frac{150\ \mathfrak{V}\ n}{R\ r\sqrt{\boxminus\ \square}}. \quad . \quad . \quad . \quad . \qquad 1$$

$$n = \frac{\mathcal{V}\ R\ r\sqrt{\boxminus\ \square}}{150\ \mathfrak{V}}. \quad . \quad . \quad . \qquad 2$$

$$\mathfrak{V} = \frac{\mathcal{V}\ R\ r\sqrt{\boxminus\ \square}}{150\ n}. \quad . \quad . \quad . \qquad 3$$

$$\boxminus\ \square = \left(\frac{150\ \mathfrak{V}\ n}{\mathcal{V}\ R\ r}\right)^2. \quad . \quad . \quad . \quad . \qquad 4$$

Example 1. A steam-engine of 3 feet diameter of cylinder and 5 feet stroke of piston is to make $n = 70$ revolutions per minute, with a boiler of $\boxminus = 164$ square feet of fire-grate and heating surface $\square = 4850$ square feet. The steam to be cut off at half stroke; feed-water 120°, for which $R = 1.087$; height of chimney 85 feet, for which $r = 1.27$. Required what steam-pressure the boiler can carry under the above conditions?

$\mathfrak{V} = 7.061 \times 5 = 35.305$ cubic feet of steam for each revolution, to which add for clearance and steamport 1.7 cubic feet, making $\mathfrak{V} = 37$ cubic feet.

$$\text{Steam volume,} \qquad \mathcal{V} = \frac{150 \times 37 \times 70}{1.087 \times 1.27\sqrt{164 \times 4850}} = 274.94.$$

Find the steam-pressure corresponding to this volume (see Steam Table, Nystrom's *Pocket-Book*), which is 82 pounds to the square inch, the pressure required.

Example 2. How many revolutions per minute may be expected from an engine using $\mathfrak{V} = 15$ cubic feet of full steam of 50 pounds to the square inch for each revolution, when the steam-boiler is $\boxminus = 84$ and $\square = 2480$ square feet, the temperature of the feed-water being 90°, for which $R = 1.054$, height of chimney 40 feet, for which $r = 0.91$?

The steam-volume for 50 lbs. is $\mathfrak{V} = 397.51$.

$$\text{Revolutions,} \quad n = \frac{397.51 \times 1.054 \times 0.91\sqrt{84 \times 2480}}{150 \times 15} = 24.46.$$

Formula 4 gives the size of steam-boiler required for a given-sized engine and revolution of the same.

Example 4. What size steam-boiler is required for an engine using $\mathfrak{V} = 20$ cubic feet of full steam of pressure 60 pounds to the square inch, to make $n = 48$ revolutions per minute; height of chimney 75 feet and temperature of feed-water 100° ?

$$⊟\,□ = \left(\frac{150 \times 20 \times 48}{384.15 \times 1.0647 \times 1.2}\right)^2 = 86086.5.$$

Suppose the heating surface in the boiler to be $□ = 25\,⊟$, then $25\,⊟^2 = 86086.5$.

$$\text{The required fire-grate,} \quad ⊟ = \sqrt{\frac{86086.5}{25}} = 58.68 \text{ square feet.}$$

$$\text{Heating surface,} \quad □ = 58.68 \times 25 = 1467 \text{ square feet.}$$

§ 59. QUANTITY OF FEED-WATER BY AREAS OF FIRE-GRATE AND HEATING SURFACE.

W = cubic feet of water to be fed into the boiler per minute.
d = diameter in inches of the pump-piston or feed-plunger.
s = stroke in inches of piston or plunger.
n = pumping strokes per minute.

$$W = \frac{\sqrt{⊟\,□}}{150}. \quad . \quad . \quad . \quad . \quad . \quad 1$$

$$\frac{0.7854\,d^2\,s\,n}{1728} = \frac{\sqrt{⊟\,□}}{150}.$$

$$d^2\,s\,n = 14.668\sqrt{⊟\,□}.$$

Add 36 per cent. for feeding the boiler with safety and allowing for slip-water.

$$d^2\,s\,n = 20\sqrt{⊟\,□}. \quad . \quad . \quad . \quad . \quad . \quad 2$$

$$d = \sqrt{\frac{20\sqrt{\boxminus\,\square}}{s\,n}}. \qquad \ldots \qquad 3$$

$$s = \frac{20\sqrt{\boxminus\,\square}}{d^2\,n}. \qquad \ldots \qquad 4$$

$$n = \frac{20\sqrt{\boxminus\,\square}}{d^2\,s}. \qquad \ldots \qquad 5$$

Example 1. How much water is required per minute to feed a boiler of $\boxminus = 45$ square feet fire-grate, and $\square = 1250$ square feet heating surface?

$$W = \frac{\sqrt{45 \times 1250}}{150} = 1.6 \text{ cubic feet.}$$

Example 3.—What diameter must be given to a feed-plunger of $s = 8$ inches stroke, making $n = 50$ strokes per minute, to feed the boiler of $\boxminus = 36$ square feet fire-grate and $\square = 1296$ square feet heating surface?

$$d = \sqrt{\frac{20\sqrt{36 \times 1296}}{8 \times 50}} = 3.3 \text{ inches.}$$

§ 60. CAPACITY OF THE FEED-PUMP BY THE SIZE OF THE STEAM-CYLINDER.

D = diameter in inches of the steam-cylinder, double acting.
S = part of the stroke in inches under which steam is fully admitted, including clearance and capacity of steamports.
$\mathfrak{V}$ = Steam-volume corresponding to the steam-pressure.
d = diameter in inches of the pump-plunger, single acting.
s = stroke of the pump-plunger in inches.

It is supposed that the feed-pump is connected with the engine, so as to make the same number of strokes per unit of time as does the steam-piston.

$$\mathfrak{V}\,d^2\,s = 2D^2\,S$$

$$s = \frac{2D^2\,S}{\mathfrak{V}\,d^2} \qquad \ldots \qquad 1$$

$$d = D\sqrt{\frac{2S}{\mathfrak{V}\,s}}. \qquad \ldots \qquad 2$$

Add 50 per cent. to the last number for safety in feeding the boiler and for slip-water. The practical formula should then be

$$\text{Diameter of plunger } d = \sqrt{\frac{3D^2S}{\rlap{/}V s}} \quad . \quad . \quad . \quad 1$$

$$\text{Stroke of plunger } s = \frac{3D^2S}{\rlap{/}V d^2} \quad . \quad . \quad . \quad 2$$

Example 1.—The diameter of a steam-cylinder is $D = 36$ inches, full steam-pressure 75 pounds, cut-off at 32 inches, to which add for clearance and capacity of steamports say 2 inches, making $S = 34$ inches. The stroke of the feed-plunger is designed to be $s = 24$ inches. Required the diameter of the plunger?

$$d = \sqrt{\frac{3 \times 36^2 \times 34}{294.61 \times 24}} = 1.8696 \text{ ; say 2 inches.}$$

RADIATION OF HEAT FROM STEAM-PIPES, BOILERS AND STEAM-CYLINDERS.

§ 61. The quantity of heat radiated from a hot surface into the air varies directly as the difference of temperature of the hot surface and of the surrounding air. The radiation per square foot is not constant for cylindrical surfaces under 12 inches in diameter, but varies in an arithmetical ratio inversely as the square of the diameter—that is, small steam-pipes radiate more heat per square foot of surface than do large ones up to 12 inches diameter. For diameters over 12 inches the quantity of heat radiated is directly as the surface exposed to free air.

The thickness of metal, within the limit of ordinary practice, does not seem to materially affect the quantity of heat radiated from uncovered surfaces.

When the radiating surface is covered with felt and canvas outside, the check of radiation of heat is greater for small diameters of pipe than for larger ones with the same thickness of covering, as will be seen in the accompanying Table XVI.

D = outside diameter of steam-pipe in inches.
L = length in feet of cylinder or pipe.
A = radiating area in square feet.
T = temperature Fahr. of the steam in the steam-pipe.
t = temperature of the external air.
h = units of heat radiated per hour.

C = cubic feet of steam of temperature T condensed per hour.

l = latent heat per cubic foot of steam of temperature T, which is denoted by L' in Steam Table, see *Pocket-Book*.

p = pressure per square inch of the steam.

HP = horse-power lost by radiation.

m = percentage of heat or power gained by covering the pipe with felt. (See Table XVI.)

n = exponent of the wind, which varies with the velocity of the current of air passing the radiating surface as follows:

Calm.	Gentle.	Brisk.	Storm.
$n=1.2$	$n=1.22$	$n=1.24$	$n=1.26$

§ 62. RADIATION FROM UNCOVERED SURFACES.

Heat radiated per hour, $$h=0.505A(T-t)^n. \quad . \quad 1$$

For cylinder or pipes over 12 inches in diameter, the radiation per hour will be

Units of heat, $$h=0.1322DL(T-t)^n. \quad . \quad 2$$

For cylinders or pipes under 12 inches in diameter, the radiation per hour will be

Units of heat, $$h=\frac{DL}{3404.8}[450+(12-D)^2](T-t)^n. \quad 3$$

The volume in cubic feet of steam condensed per hour will be

$$C=\frac{h}{l}. \quad . \quad . \quad . \quad . \quad . \quad 4$$

Horse-power lost by radiation of h units of heat per hour will be

$$\text{HP}=\frac{Cp}{13748.4}. \quad . \quad . \quad . \quad . \quad 5$$

Example 1. How many units of heat are radiated per hour from an uncovered steam-boiler exposing $A=198$ square feet of radiating surface in a gentle breeze of $t=45°$, when the steam-pressure in the boiler is $p=65$ pounds to the square inch?

Units of heat, $$h=0.505\times198(311.86-45)^{1.22}=91206.$$

$$311.86 - 45 = 266.86.$$

Logarithm,	266.86 = 2.4262835	
Multiply by exponent,	1.22	
	48525670	
	48525670	
	24262835	
	912.15 = 2.960065870	
Add log.	198 = 2.2966652	
Add log.	0.505 = 0.7032914 − 1	
Units of heat,	91206 = 4.9600225	

Example 4. How many cubic feet of steam are condensed by the radiation of $h = 91206$ units of heat per hour? Latent heat, $l = 170$.

$$C = \frac{91206}{170} = 536.5 \text{ cubic feet.}$$

	4.9600225
Subtract log.	170 = 2.2304489
Cubic feet of steam,	536.5 = 2.7295736

Example 5. How much horse-power is lost by the radiation in the preceding examples? $C = 407.48$ cubic feet, and $p = 65$ pounds

$$\text{Power lost,} \quad \text{IP} = \frac{536.5 \times 65}{13748.4} = 2.5375 \text{ horse-power.}$$

	log. 536.5 = 2.7295736	add
	log. 65 = 1.8129134	
	4.5424870	
Subtract	log. 13748.4 = 4.1382521	
Horse-power lost,	2.5375 = 0.4042349	

Example 2. An uncovered steam-pipe is $D = 8$ inches diameter and $L = 28$ feet long, conducting steam of $p = 80$ pounds pressure, and temperature $T = 324°$. The temperature of the surrounding air is $t = 40°$ of brisk wind. Required the units of heat lost, the cubic feet of steam condensed per hour and the horse-power lost by radiation from the pipe?

$$\text{Units of heat,} \quad h = \frac{8 \times 28}{3404.8}[450 + (12 - 8)^2](324 - 40)^{1.24} = 33780.$$

The whole calculation is practically set up as follows by logarithms:

		Logarithms.
	$324 - 40 = 284 =$	2.4533183
Multiply by exponent, ...		1.24
		98132732
		49066366
		24533183
	$(324 - 40)^{1.24} =$	+ 3.042114692
	$(12 - 8)^2 = 16 + 450 = 466 =$	+ 2.6683859
	$8 \times 28 = 224 =$	+ 2.3502480
	$8 \times 28[450 + (12 - 8)^2](324 - 40)^{1.24} =$	+ 8.0607486
Coefficient, ...	$3404.8 =$	− 3.5320916
The required units of heat, . .	$h = 33780 =$	+ 4.0286570
Latent heat per cubic foot, . .	$l = 196.84 =$	− 2.2943339
Cubic feet of steam condensed, .	$C = 171.52 =$	+ 2.2343231
Steam-pressure, ...	$p = 80 =$	+ 1.9030900
	$Cp =$	+ 4.1374131
Coefficient, ...	$13748.4 =$	− 4.1382521
Horse-power lost, . .	$\text{HP} = 0.99807 =$	+ 0.9991610 − 1

Say one horse-power lost by radiation.

It is supposed in this example that the steam is working a high-pressure engine without expansion. For a condensing engine take the steam-pressure above vacuum and multiply the lost power by $1 + hyp.log.$ of the expansion, and the product will be the correct horse-power lost.

COVERED STEAM-PIPES.

§ 63. When the steam-pipe is covered with felt and canvas outside, there is very little heat radiated, as will be seen in the accompanying table, which gives the heat and power saved by covering of different thickness.

Suppose the loss by radiation of heat from an uncovered steam-pipe 6 inches in diameter is $\text{HP} = 2$ horse-power; then, by covering the pipe with felt one inch thick will save 86 per cent. of the 2 horse-power, or $2 \times 0.86 = 1.72$ horse-power, and the loss by radiation from the covered pipe will be only $2 - 1.72 = 0.28$ of a horse-power.

TABLE XVI.

Percentage *m* of Heat or Power Gained by Covering Steam-pipes with Felt and Canvas Outside.

Diam. pipe.	Thickness in Inches of Felt Covering.									
	$\frac{1}{4}$	$\frac{3}{8}$	$\frac{1}{2}$	$\frac{3}{4}$	1	$1\frac{1}{2}$	2	3	4	6
D	m	m	m	m	m	m	m	m	m	m
1	65	76	81	86	92	94	96	98	99	100
2	63	74	80	85	90	93	95	97	98	99
3	61	72	79	84	89	92	95	96	98	99
4	59	71	77	83	88	92	94	96	97	99
5	57	69	76	82	87	91	94	96	97	99
6	54	67	74	81	86	91	94	95	97	99
7	52	66	73	81	85	90	93	95	97	99
8	50	64	71	80	85	90	93	95	97	99
9	47	62	70	79	84	89	93	95	97	99
10	45	61	69	78	84	89	92	95	96	98
11	42	59	67	78	83	88	92	94	96	98
12	40	58	66	77	83	88	92	94	96	98

STEAM-BOILER EXPLOSIONS.

§ 64. Steam-boiler explosions are caused by suddenly liberating all the work stored in the boiler.

The *work* K is the product of the three simple physical elements *force* F, *velocity* V and *time* T.

$$\textit{Work}, \ K = FVT. \qquad . \quad . \quad . \quad 1$$

The force of this work is, therefore,

$$F = \frac{K}{VT}. \qquad . \quad . \quad . \quad . \quad 2$$

When the steam-pressure in any part of the boiler is suddenly removed by bursting of the shell, the entire work of the heat stored in the steam and water is at the same time started with a velocity proportionate to the removed pressure.

When the pressure is suddenly lowered below that due to the temperature of the water, the heat in it generates steam, which raises the water bodily in the form of foam, striking the steam-side of the boiler, and the work is thus suddenly arrested. If the time of arresting the work is infinitely small, the force will, according to Formula 2, be infinitely great, and thus the boiler explodes.

§ 65. Let Fig. 3 represent the steam-boiler, consisting of a cylindrical tube of one square foot section and of indefinite length. The lower end of the tube is closed and contains one cubic foot of water, from which steam has been generated by the heat of the lamp L, and has raised the piston with the weight Q a space S from the surface of the water.

Fig. 3.

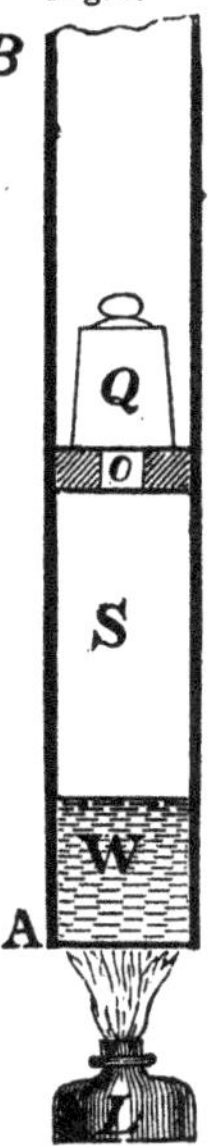

Assume the steam-pressure to be $P = 65$ pounds to the square inch above vacuum, and one cubic foot of steam between the piston and the water. Then,

In one cubic foot of water,	$H = 15485$	units of heat.
In one cubic foot of steam,	$H' = 184$	" "
Total heat in the boiler,	$H + H' = 15669$	units.

Take away the lamp, so that no more heat enters into the boiler.

Diminish gradually the weight Q; the expansion of the steam will then raise the piston, and the heat in the water will evaporate more steam until the temperature corresponds with the reduced pressure. The temperature of the water at $P = 65$ is $T = 297.84°$; and if the weight Q is gradually reduced to 14.7 pounds to the square inch on the piston, the temperature of the steam and water will be 212° Fahr.

One cubic foot of water at $T = 287.84°$ weighs 57.687 pounds, of 268.39 units of heat per pound.

§ 66. At the temperature 212° the units of heat per pound of water are 180.9 and per pound of steam 1146.6. The question now is, How many pounds of water w and how many pounds of steam s of temperature 212° are there in the boiler?

$$180.9\,w + 1146.6\,s = 15485 \text{ units of heat.}$$

$$w + s = 57.85 \text{ pounds.}$$

$$w = 57.69 - s. \quad \text{Then, } 180.9\,(57.69 - s) + 1146.6\,s = 15485.$$

Complete the calculation, which will give

$s = 5.228$ pounds of steam of . . .	5994.8	units of heat.
$w = 52.46$ pounds of water of . . .	9490.0	" "
For one cubic foot of steam add . .	184	" "
Total	15658.8	" "
The original heat was	15669.	" "
52.46 pounds of water at 212° = .	0.8767	cubic feet.
5.228 pounds of steam at 212° = .	135.58	" "
Add one cubic foot expanded four times	4	" "
Total volume of steam . .	139.58	" "

That is to say, the piston has moved $139.58 - 1.12 = 138.46$ feet from the position occupied when the weight Q was first diminished.

The work accomplished by this operation is determined as follows:

5.228 pounds of steam of pressure $P = 65 = 35.7$ cubic feet.

$65 : 14.7 = 4.47$ the expansion of the steam.

Hyperbolic log. $4.47 \doteq 1.49734$.

Work $K = 144 \times 65 \times 35.7 \times 1.4973 = 500330$ foot-pounds.

From this subtract the work of the atmosphere, which is

$k = 144 \times 14.7 \times 138.46 = 293100$ foot-pounds.

Then $500330 - 293100 = 207230$ foot-pounds of work done against the atmosphere.

Divide this work by 550 times the number of seconds occupied in its execution, and the quotient will be the horse-power of the operation.

§ 67. Now suppose the piston to be firmly fixed in the position shown by the illustration Fig. 3, and instead of gradually diminishing the weight Q, let it be suddenly removed, leaving the hole o open for the steam to escape. The moment the steam-pressure on the surface of the water is removed or reduced, the heat will generate steam of a pressure of 65 pounds to the square inch in all parts of the water; and as there is not a corresponding pressure on its surface, the steam will lift the water bodily in the form of foam, striking the immovable piston, and thus explode the boiler.

Under the conditions assumed, the work of this explosion will be 911160 foot-pounds, accomplished, no doubt, within the time of one second, in which case $207230 : 550 = 1337$ horse-power of the explosion of only one cubic foot of water, of which only $1 - 0.8767 = 0.1233$ of that cubic foot was converted into steam.

The mystery of steam-boiler explosions is thus explained.

§ 68. The investigation becomes more simple by way of algebraical formulas, for which letters will denote—

W = pounds of water under steam-pressure in the boiler before explosion.

w = pounds of water reduced to temperature 212°, and not evaporated in the explosion.

lbs. = pounds of water evaporated to steam in the explosion and expanded to the pressure of the atmosphere.

h = units of heat per pound of water in the boiler before explosion.

P = steam-pressure in pounds per square inch above vacuum in the boiler before explosion.

C = cubic feet of steam of atmospheric pressure generated by the heat in the water before explosion.

K = destructive work of the explosion in foot-pounds.

$$\text{Units of heat } Wh = 181\, w + 1147 \text{ lbs.} \quad . \quad . \quad 3$$

$$W = w + \text{lbs. and } w = W - \text{lbs.} \quad . \quad . \quad . \quad . \quad 4$$

$$Wh = 181\, (W - \text{lbs.}) + 1147 \text{ lbs.} \quad . \quad . \quad . \quad 5$$

$$\text{lbs.} = \frac{W}{966}(h - 181) \quad . \quad . \quad . \quad . \quad 6$$

The weight per cubic foot of steam of atmospheric pressure is 0.038, and the volume of steam evaporated and expanded in the explosion to atmospheric pressure will be $996 \times 0.038 = 36.7$.

$$C = \frac{W}{36.7}(h - 181) \quad . \quad . \quad . \quad . \quad 7$$

The volume of this steam under the pressure P was

$$\frac{14.7\, C}{P - 14.7} \quad . \quad . \quad . \quad . \quad 8$$

The gross work done by the explosion will then be

$$k = \frac{144 \times 14.7 P C}{P - 14.7}\, hyp.log. \frac{P}{14.7} \quad . \quad . \quad 9$$

From this work should be subtracted the reaction of the atmosphere, which is $144 \times 14.7\, C$.

The remainder will be the destructive work of the explosion, namely,

$$k = 2116.8\, C\left(\frac{P}{P - 14.7}\, hyp.log. \frac{P}{14.7} - 1\right). \quad . \quad 10$$

Example 7.—A steam-boiler containing 125 cubic feet of water explodes under a steam-pressure of $P = 85$ pounds to the square inch. Required the destructive work of the explosion?

Under this pressure the temperature of the water is 316.08°, and weighs 57.21 pounds per cubic foot.

$$W = 125 \times 57.21 = 7151.25 \text{ pounds.}$$

The steam-volume generated by the explosion is

$$C = \frac{7151.25}{36.7}(287 - 181) = 20655 \text{ cubic feet.}$$

$$K = 2116.8 \times 20655\left(\frac{85}{85 - 14.7} hyp.log. \frac{85}{14.7} - 1\right) = 49200550 \text{ foot-pounds,}$$

the required work of destruction.

This work is equivalent to that of the explosion of 246 pounds of gunpowder, which is more than double the work of a charge from a 20-inch gun. A great part of the work of steam-boiler explosions is consumed in setting the air into vibration, which makes the report.

§ 69. A laborer working 8 hours per day with a power of 50 effect accomplishes a work of 1,440,000 foot-pounds of work, called "workmanday."

The work of the above steam-boiler explosion 49200550 : 1440000 = 34 workmandays. It would require 34 men to work one day, or one man 34 days, to do the same amount of work.

The work of the steam in the boiler prior to the explosion is not included in the preceding formulas and examples, because it is an insignificant quantity compared with that of the heat in the water. The bursting of a vessel full of steam without water will cause very little damage compared with that of a vessel full of water under steam-pressure.

c = cubic feet of steam in a boiler of
P = pressure per square inch above vacuum.
k = work of explosion of the steam only.

$$k = 144c\left(P hyp.log. \frac{P}{14.7} + 14.7 - P\right) \quad . \quad . \quad . \quad 11$$

CAUSE AND PREVENTION OF STEAM-BOILER EXPLOSIONS.

§ 70. The bursting of a steam-boiler is a preliminary process to the explosion.

In a vessel composed of any non-elastic material and filled with water hermetically sealed in it, if that water is frozen solid, the expansion of the ice will most likely burst the vessel, but there will be no explosion, because there is no explosive agency in it.

A steam-boiler full of cold water and tested with hydrostatic pressure until it bursts, will not explode; but if that cold water is heated to a temperature corresponding to the bursting pressure, there will be an explosion.

The iron in steam-boilers, like any other material subjected to bursting strain, breaks at the weakest point; but it is difficult to find the location of that point, and very often boilers are not constructed, inspected or managed with sufficient care to guard against bursting. Thus steam-boiler explosions are caused by various neglects in guarding against such accidents—namely,

First. By long use boilers become weakened by corrosion, which acts unevenly on different kinds of iron and in different parts of the boiler, and if not properly inspected and the weakened places repaired, the boiler may burst and explode.

Second. The general construction, with staying and bracing of steam-boilers, is often very carelessly executed, and results in explosion. This kind of explosions are often indicated long before the accident occurs, by leakage of the boiler; when the engineer, not suspecting the approaching danger, limits the remedies generally to efforts toward stopping the leak. Leakage from bad caulking or packing is easily distinguished from that of bad or insufficient bracing, in which latter case the fire ought to be hauled out, the steam blown off gradually, and the boiler secured with proper bracing.

Third. The strength and quality of iron in the original construction are not always properly selected to correspond with the duty expected of the boiler, which neglect causes explosion.

Fourth. Single-riveted joints weaken the strength of a boiler about 50 per cent. of that of the solid plate, and boilers therefore often burst by tearing the plate between the rivets. This defect can be remedied by making double-riveted joints, which, if properly proportioned, are (by experiments) as strong as the solid plate.

Fifth. Explosion is sometimes caused from low water in the boiler, but more rarely than is generally supposed. When the fire crown and flues are subjected to a strong heat and not covered with water, the steam does not absorb the heat fast enough to prevent the iron from becoming so hot that it cannot withstand the pressure, but collapses from weakness, and the boiler explodes. There are several good inventions for preventing too low water in boilers, which should invariably be used.

Sixth. Steam-boilers often burst from strain in uneven expansion or shrinkage of the iron by sudden change of temperature. When the fire is too quickly lighted or extinguished, there is not time enough for the heat to communicate alike to and from all parts of the boiler, the effect of which has often been the cause of bursting the boiler. When cold feed-water is injected near to the fire-place, it absorbs the heat quickly and cools that part of the heating surface, and when the feed is not evenly supplied, but alternately stopped and forced in with the full capacity of the pump, there will be a corresponding contraction and expansion of that part of the iron, the work of which is injurious to and may finally cause the bursting of the boiler. The feed-water should be heated to at least 100° for condensing engines and 180° for high-pressure engines, and injected at some distance from the furnace.

Seventh. It is a very bad practice to make boiler-ends of cast-iron, composed of a flat disc of from two to three inches thick, with a flange of from one to two inches thick, with cast rivet holes. The first shrinkage in the cooling of such a plate causes a great strain, which is increased by riveting the boiler to it. Any sudden change of temperature in such plate, either by starting or putting out the fire, might crack the plate and cause explosion of the boiler.

Such accidents can be avoided by making the boiler-ends of wrought-iron plates properly stayed or made concave on the steam side.

Eight. In cold weather, when the boilers have been at rest for some time, the water in them may be frozen to ice; then, when fire is quickly made in them, some parts are suddenly heated and expand, whilst other parts still remain cold, thus causing an undue strain which may so injure the boiler that it will not be able to bear the required steam-pressure, and explosion follows.

Such accident can be avoided by a slow and cautious firing, so that all the ice may be thoroughly melted before steam is generated in any part of the boiler.

Ninth. When a number of boilers are placed close together and connected to a common steam-pipe, the weakest part in either one of them is the measure of safety for all the rest; for however strong the other boilers may be, when the weakest one bursts all the rest will most likely explode simultaneously, as has often been the case.

Tenth. Steam-boiler explosions are thus not always caused by the pressure of steam alone, but most frequently by the expansion and contraction of the iron composing the boilers. A steam-boiler which is perfectly safe with a working pressure of 200 pounds may explode with a pressure of 20 pounds to the square inch.

Eleventh. See "Superheating Steam" for another possible cause of explosions.

STRENGTH AND SAFETY OF STEAM-BOILERS.

§ 71. The law in the United States regulating the strength and safety of steam-boilers, passed by Congress February 28, 1871, and enforced February 28, 1872, is that all the plates used in steam-boilers shall be stamped with the number of pounds equal to the breaking-strength per square inch section of the iron. One-sixth of the stamped number is taken as the safety or working strength of the iron in the boiler.

The law requires that steam-boilers must be tested with hydrostatic pressure of 50 per cent. above the working pressure allowed.

The following quotations are copied from the rules prescribed for the Boiler Inspectors:

"Where flat surfaces exist, the inspector must satisfy himself that the bracing, and all other parts of the boiler, are of equal strength with the shell, and he must also, after applying the hydrostatic test, thoroughly examine every part of the boiler to see that no weakness or fracture has been caused thereby. Inspectors must see that the flues are of proper thickness to avoid the danger of collapse. Flues of sixteen inches in diameter must not be less than one-quarter of an inch in thickness, and in proportion for flues of a greater or less diameter."

"Every iron or steel plate intended for the construction of boilers to be used on steam-vessels shall be stamped by the manufacturer in the following manner, viz.: At the diagonal corners, at a distance of about four inches from the edges, and also at or near the centre of the plate, with the name of the manufacturer, the place where manufactured, and the number of pounds tensile strain it will bear to the sectional square inch."

"The manner of inspecting, testing and stamping boiler-plates, by the United States inspectors, shall be as follows, viz.:

"The sheets to be inspected and tested shall be selected by the inspectors, indiscriminately, from the lot presented, and shall not be less than one-tenth of the entire lot so presented, and from every such selected sheet the inspector shall cause a piece to be taken, for the purpose of ascertaining its strength, the area of which shall equal one-quarter of one square inch, and the force at which this piece can be parted in the direction of its fibre or grain, represented by pounds avoirdupois multiplied by four, shall be the tensile strength, and the lot from which the test-sheets were taken shall not be marked above the lowest number represented by these tests. The inspector shall also subject a piece taken from each selected sheet to repeated heating and cooling, and shall bend it short, both in a hot and a cold state, and shall draw it out under the hammer, as it is called, in order to ascertain the other qualities mentioned in Section 36 of the act aforesaid; and should these test-pieces be found deficient in these qualities, the inspectors shall refuse to place the government stamp on the lot from which these test-sheets were taken; but if the test-pieces should prove to possess these qualities, then the inspector shall proceed to stamp the entire lot from which they were taken with the letters 'U.S.' and the figure denoting the inspection-district in which the inspection was made."

"All boiler-plates tested and stamped as above shall be considered as having been inspected according to law; but should any local or other inspector have valid reasons for believing that fraud has been practiced, and that the stamps upon any such boiler-plates are false, in whole or in part, he is empowered to re-inspect and test the same."

"The provisions of this rule shall take effect as soon as the inspectors are appointed, and the manufacturers of boiler-plates notified of the same."

The rule for proportioning the strength of boilers to the steam-pressure is as follows:

Rule. "Multiply one-sixth ($\frac{1}{6}$) of the lowest tensile strength found stamped on any plate in the cylindrical shell by the thickness expressed in parts of an inch of the thinnest plate in the same cylindrical shell, and divide the product by the radius or half the diameter of the shell expressed in inches, and the quotient will be the steam-pressure in pounds per square inch allowable in single-riveted boilers, to which add twenty per centum for double riveting."

No allowance is made by this rule for the metal punched away by the holes in the plate. Allowing 66 per cent. of metal between the holes, the safety strength will be one-quarter of the ultimate strength.

The rule is more simply expressed by algebraical formulas, as follows:

S = breaking-strain in pounds per square inch, stamped on the boiler-plate.

t = thickness of the plate in fractions of an inch.

D = inside diameter of the boiler in inches.

p = steam-pressure in pounds per square inch allowable in the boiler, single riveted.

§ 72. **Safety Strength of Single-Riveted Joints.**

Steam-pressure, $p = \frac{S\,t}{3\,D}$. 1

Diameter of boiler, $D = \frac{S\,t}{3\,p}$. 2

Thickness of plate, $t = \frac{3\,D\,p}{S}$. 3

Breaking-strain, $S = \frac{3\,D\,p}{t}$. 4

Example 1. A steam-boiler of $D = 48$ inches diameter and thickness of plates $t = 0.375$ of an inch is stamped with a breaking-strain $S = 55{,}000$ pounds. Required the steam-pressure the boiler is allowed to carry?

$p = \frac{55000 \times 0.375}{3 \times 48} = 143.2$ pounds to the square inch for single-riveted joints.

For double-riveted joints $143.2 \times 1.2 = 171.8$ pounds to the square inch.

§ 73. **Safety Strength of Double-riveted Joints.**

Steam-pressure, $p = \frac{0.4\,S\,t}{D}$. 5

Diameter of boiler, $D = \frac{0.4\,S\,t}{p}$. 6

Thickness of plate, $t = \frac{D\,p}{0.4\,S}$. 7

Breaking-strain, $S = \frac{D\,p}{0.4\,t}$. 8

Example 8. A double-riveted boiler is to be constructed to carry $p = 80$ pounds of steam in a diameter $D = 96$ inches, with $t = 0.3$ of an inch thickness of plate. Required the stamp on the plates?

$$S = \frac{96 \times 80}{0.4 \times 0.3} = 64{,}000 \text{ stamp.}$$

The following tables are calculated from the above formulas for single and double-riveted boilers.

TABLE XVII.

Boiler Plates Stamped 45,000 lbs. Safety-strain $\frac{1}{6}$ = 7500.

Diameter of boiler, inches.	Thickness of boiler-plate in fractions of an inch.									
	$\frac{3}{16}$ = 0.1875 Riveting.		$\frac{1}{4}$ = 0.25 Riveting.		$\frac{9}{32}$ = 0.28125 Riveting.		$\frac{5}{16}$ = 0.3125 Riveting.		$\frac{11}{32}$ = 0.34375 Riveting.	
	Single.	Double.	Single.	Double.	Single.	Double.	Single.	Double.	Single.	Double.
D	Pressures.		Pressures.		Pressures.		Pressures.		Pressures.	
36	78.12	93.74	104.2	125.	117.2	140.6	130.2	156.2	143.2	171.8
38	74.	88.8	98.6	118.3	110.9	133.1	123.3	148.	135.6	162.8
40	70.31	84.37	93.7	112.4	105.4	126.5	117.2	140.6	128.1	154.7
42	66.96	80.35	89.2	107.	100.4	120.5	111.6	133.9	122.7	147.3
44	63.92	86.7	85.2	102.2	95.85	115.	106.5	127.8	117.1	140.5
48	58.59	70.3	78.1	93.72	82.87	99.45	97.65	117.2	107.4	128.9
54	52.	62.4	69.44	83.32	78.12	93.74	86.8	104.2	95.5	114.6
60	46.87	56.24	62.5	75.	70.31	84.37	78.12	93.74	85.93	103.1
66	42.79	51.34	56.86	68.17	63.93	76.71	71.	85.2	78.1	93.72
72	39.	46.8	52.	62.4	58.55	70.26	65.1	78.12	71.61	85.93
78	36.	43.	49.34	58.86	54.67	65.6	60.	72.1	66.05	79.26
84	33.48	40.17	44.64	53.56	50.22	60.26	55.8	66.96	61.38	73.65
90	31.25	37.5	41.66	50.	46.83	56.19	52.	62.5	57.25	68.7
96	29.28	35.53	39.	46.8	43.91	52.69	48.82	58.58	53.7	64.44
102	27.56	33.07	36.76	44.11	41.35	49.62	45.95	55.14	50.53	60.64
108	26.	31.2	34.72	41.86	39.06	46.87	43.4	52.1	47.75	57.3
120	23.43	28.12	31.25	37.5	35.15	42.18	39.06	46.87	42.96	51.56
D	$\frac{3}{8}$ = 0.375		$\frac{7}{16}$ = 0.4375		$\frac{1}{2}$ = 0.5		$\frac{9}{16}$ = 0.5625		$\frac{5}{8}$ = 0.625	
36	156.2	187.5	182.3	218.8	208.3	250.	234.3	281.2	260.4	312.5
38	148.	177.6	172.6	207.1	197.2	236.6	221.8	266.2	246.6	296.
40	140.6	168.7	164.	196.8	187.4	224.9	210.8	253.	234.4	281.2
42	133.9	160.7	156.1	187.4	178.4	214.	200.8	241.	223.2	267.8
44	127.8	153.4	148.9	178.7	170.	204.5	191.7	230.	213.	255.6
48	117.2	140.6	136.7	164.	156.2	187.4	165.7	198.9	195.3	234.4
54	104.2	125.	121.5	145.8	138.9	166.6	156.2	187.5	173.6	208.4
60	93.75	112.5	109.4	131.1	125.	150.	140.6	168.7	156.2	187.5
66	85.2	102.2	99.45	119.3	113.7	136.3	127.9	153.4	142.	170.4
72	78.12	93.74	91.06	109.3	104.	124.8	117.1	140.5	130.2	156.2
78	72.1	86.53	85.39	102.4	98.68	117.7	109.3	131.2	120.	144.2
84	66.96	80.35	78.12	93.74	89.28	107.1	100.4	120.5	111.6	133.9
90	62.5	75.	72.91	87.5	83.33	100.	93.7	112.4	104.	125.
96	58.58	70.29	68.29	81.95	78.	93.6	87.8	105.4	97.6	117.2
102	55.12	66.14	64.32	77.19	73.53	88.22	82.7	99.2	91.9	110.3
108	52.1	62.5	60.77	72.93	69.45	83.3	78.1	93.7	86.8	104.2
120	46.87	56.25	54.68	65.62	62.5	75.	70.3	84.3	78.1	93.7

TABLE XVIII.

Boiler Plates Stamped 50,000 lbs. Safety-strain $\frac{1}{6}$ = 8333.3.

Diameter of boiler, inches.	Thickness of boiler-plate in fractions of an inch.									
	$\frac{3}{16}$ = 0.1875 Riveting.		$\frac{1}{4}$ = 0.25 Riveting.		$\frac{9}{32}$ = 0.28125 Riveting.		$\frac{5}{16}$ = 0.3125 Riveting.		$\frac{11}{32}$ = 0.34375 Riveting.	
	Single.	Double.	Single.	Double.	Single.	Double.	Single.	Double.	Single.	Double.
D	Pressures.		Pressures.		Pressures.		Pressures.		Pressures.	
36	86.8	104.2	115.7	138.9	130.2	156.2	144.7	173.6	159.1	191.
38	82.23	98.68	109.6	131.5	123.3	148.	137.	164.5	150.7	180.8
40	78.12	93.74	104.1	125.	117.1	140.6	130.2	156.2	143.2	171.8
42	74.49	89.38	99.2	119.	111.6	133.9	124.	148.8	136.4	163.7
44	71.	85.2	94.69	113.6	106.5	127.7	118.4	142.	130.2	156.2
48	65.1	78.12	86.8	104.1	97.4	116.9	108.	130.2	119.1	142.9
54	57.62	69.44	77.16	92.59	86.8	104.1	96.45	115.7	101.	121.3
60	52.	62.4	69.44	83.33	78.12	93.74	86.8	104.1	95.45	114.5
66	47.34	56.8	63.13	75.75	71.02	85.22	78.91	94.69	86.8	104.1
72	43.4	52.	57.87	69.44	65.11	78.13	72.35	86.8	79.57	95.48
78	40.	48.	53.67	64.4	60.22	72.26	66.77	80.12	73.44	88.13
84	37.2	44.64	49.6	59.5	55.8	66.96	62.	74.4	68.2	81.84
90	34.72	41.66	46.29	55.55	52.08	62.5	57.87	69.44	63.65	76.38
96	32.55	39.	43.4	52.	48.82	58.59	54.25	65.1	59.67	71.61
102	30.63	36.77	40.66	48.79	45.87	55.04	51.08	61.29	56.17	67.41
108	28.81	34.72	38.58	46.29	43.4	52.08	48.22	57.85	53.03	63.63
120	26.	31.2	34.72	41.66	39.06	46.87	43.4	52.08	47.74	57.29
D	$\frac{3}{8}$ = 0.375		$\frac{7}{16}$ = 0.4375		$\frac{1}{2}$ = 0.5		$\frac{9}{16}$ = 0.5625		$\frac{5}{8}$ = 0.625	
36	173.6	208.3	202.5	243.	231.5	277.8	260.4	312.4	289.4	347.2
38	164.4	197.3	191.8	230.2	219.3	263.1	246.6	296.	274.	329.
40	156.2	187.5	182.2	218.7	208.3	250.	234.2	281.2	260.4	312.4
42	148.8	178.6	173.6	208.3	198.4	238.	223.2	267.8	248.	297.6
44	142.	170.4	165.7	198.8	189.4	227.3	213.	255.4	236.8	284.
48	130.2	156.2	151.9	182.3	173.6	208.3	194.8	233.8	216.	260.4
54	115.7	138.9	135.	162.	154.3	185.2	173.6	208.2	192.9	231.4
60	104.1	125.	121.5	145.8	138.9	166.6	156.2	187.5	173.6	208.2
66	94.69	113.6	110.4	132.5	126.2	151.5	142.	170.4	157.8	189.4
72	86.8	104.1	101.2	121.5	115.7	138.9	130.2	156.2	144.7	173.6
78	80.12	96.15	93.71	112.4	107.3	128.8	120.4	144.5	133.5	160.2
84	74.4	89.28	86.8	104.1	99.2	119.	111.6	133.9	124.	148.8
90	69.44	83.	81.01	97.21	92.58	111.1	104.1	125.	115.7	138.9
96	65.1	78.2	75.95	91.14	86.8	104.	97.64	117.2	108.5	130.2
102	61.27	73.54	71.29	85.55	81.32	97.58	91.74	110.1	102.1	122.6
108	57.85	69.45	67.5	81.	77.15	92.6	86.8	104.1	96.44	115.7
120	52.08	62.49	60.76	72.92	69.44	83.33	78.12	93.74	86.8	104.1

TABLE XIX.

Boiler Plates Stamped 55,000 lbs. Safety-strain $\frac{1}{6}$ = 9166.6.

Diameter of boiler, inches.	Thickness of boiler-plate in fractions of an inch.									
	$\frac{3}{16}$ = 0.1875 Riveting.		$\frac{1}{4}$ = 0.25 Riveting.		$\frac{9}{32}$ = 0.28125 Riveting.		$\frac{5}{16}$ = 0.3125 Riveting.		$\frac{11}{32}$ = 0.34375 Riveting.	
	Single.	Double.	Single.	Double.	Single.	Double.	Single.	Double.	Single.	Double.
D	Pressures.		Pressures.		Pressures.		Pressures.		Pressures.	
36	95.48	114.6	127.3	152.8	143.2	171.8	159.1	190.9	175.	210.
38	90.46	108.5	120.6	144.7	135.6	162.7	150.7	180.9	165.8	198.9
40	85.93	103.1	114.6	137.5	128.9	154.7	143.2	171.9	157.5	189.
42	81.84	98.2	109.1	130.9	122.7	147.3	136.4	163.7	150.	180.1
44	78.12	93.74	104.1	125.	117.1	140.6	130.2	156.2	143.2	171.8
48	71.61	85.93	95.43	114.6	107.4	128.8	119.3	143.2	131.2	157.5
54	63.65	76.38	84.87	101.8	95.4	114.5	106.	127.3	116.6	140.
60	57.29	68.74	76.38	91.65	85.93	103.1	95.48	114.6	105.	126.
66	52.	62.4	69.44	83.32	78.12	93.74	86.8	104.1	95.45	114.5
72	47.74	57.28	63.65	76.38	71.6	85.92	79.56	95.48	87.52	105.
78	44.	52.8	58.76	70.51	66.1	79.32	73.45	88.13	80.79	96.95
84	40.92	49.1	54.56	65.47	61.38	73.65	68.2	81.84	75.02	90.02
90	38.19	45.82	50.92	61.1	57.28	68.73	63.65	76.38	70.01	84.02
96	35.8	42.96	47.74	57.28	53.7	64.44	59.67	71.61	65.64	78.77
102	33.7	40.44	44.93	53.9	50.54	60.65	56.16	67.39	61.78	74.13
108	31.82	38.19	42.43	50.9	47.71	57.26	53.	63.65	58.32	69.99
120	28.64	34.37	38.19	45.82	42.96	50.56	47.74	57.29	52.51	63.02
D	$\frac{3}{8}$ = 0.375		$\frac{7}{16}$ = 0.4375		$\frac{1}{2}$ = 0.5		$\frac{9}{16}$ = 0.5625		$\frac{5}{8}$ = 0.625	
36	190.9	229.1	222.7	267.3	254.6	305.5	286.4	343.6	318.2	381.8
38	180.9	217.	217.	253.2	241.2	289.4	271.2	325.4	301.4	361.8
40	171.9	206.2	200.	240.	229.1	275.	257.8	309.4	286.4	343.8
42	163.7	196.4	190.9	229.1	218.2	261.9	245.4	294.6	272.8	327.4
44	156.2	187.5	182.2	218.6	208.3	250.	234.2	281.2	260.4	312.4
48	143.2	171.8	167.1	199.5	190.9	229.1	214.8	257.6	238.6	286.4
54	127.3	152.7	148.5	178.2	169.7	203.7	190.8	229.	212.	254.6
60	114.6	137.5	133.7	160.4	152.7	183.3	171.8	206.2	190.9	229.2
66	104.1	125.	121.4	145.7	138.9	166.6	156.2	187.5	173.6	208.2
72	95.48	114.5	111.4	133.6	127.3	152.7	143.2	171.8	159.1	190.9
78	88.13	105.7	102.7	123.3	117.5	141.	132.2	158.6	146.9	176.2
84	81.84	98 2	95.48	114.6	109.1	130.9	122.7	147.3	136.4	163.7
90	76.38	91.65	89.11	106.9	101.8	122.2	114.5	137.4	127.3	152.7
96	71.61	85.93	83.54	100.2	95.48	114.5	107.4	128.9	119.3	143.2
102	67.4	80.88	78.33	94.	89.87	107.8	101.1	121.3	112.3	134.7
108	63.65	76.35	74.25	89.1	84.85	101.8	95.42	114.5	106.	127.3
120	57.29	68.74	66.83	80.2	76.38	91.64	89.92	101.1	95.5	114.6

TABLE XX.

Boiler Plates Stamped 60,000 lbs. Safety-strain $\frac{1}{6}$ = 10,000.

Diameter of boiler, inches.	Thickness of boiler-plate in fractions of an inch.									
	$\frac{3}{16}$ = 0.1875 Riveting.		$\frac{1}{4}$ = 0.25 Riveting.		$\frac{9}{32}$ = 0.28125 Riveting.		$\frac{5}{16}$ = 0.3125 Riveting.		$\frac{11}{32}$ = 0.34375 Riveting.	
	Single.	Double.	Single.	Double.	Single.	Double.	Single.	Double.	Single.	Double.
D	Pressures.		Pressures.		Pressures.		Pressures.		Pressures.	
36	104.1	125.	138.9	166.6	156.2	187.5	173.6	208.3	190.9	229.1
38	98.68	118.4	131.6	157.9	148.	177.6	164.5	197.3	180.9	217.1
40	93.74	112.5	125.	150.	140.7	168.9	156.2	187.4	166.8	200.1
42	89.28	107.1	119.	142.8	133.8	160.6	148.7	178.6	163.6	196.4
44	85.22	102.2	113.6	136.3	127.8	153.3	142.	170.4	156.2	187.4
48	78.12	93.74	104.1	125.	117.1	140.6	130.2	156.2	143.2	171.8
54	69.44	82.44	92.59	110.1	104.1	125.	115.7	138.9	127.3	152.7
60	62.4	75.	83.33	100.	93.71	113.4	104.1	125.	114.5	137.4
66	56.8	68.1	75.75	90.9	85.22	102.2	94.69	113.6	104.1	125.
72	52.	62.4	69.44	83.32	78.12	93.74	86.8	104.1	95.45	114.5
78	48.	57.6	64.4	76.92	72.26	86.71	80.12	96.15	88.13	105.7
84	44.64	53.52	59.5	71.4	66.95	80.34	74.4	89.28	81.84	98.21
90	41.66	50.	55.55	66.66	62.49	75.	69.44	83.33	76.38	91.66
96	39.	46.8	52.	62.4	58.55	70.26	65.1	78.12	71.61	85.93
102	36.76	44.12	49.02	58.8	55.14	66.17	61.27	73.51	67.4	80.88
108	34.72	41.22	46.29	55.05	52.07	62.48	57.85	69.45	63.65	76.38
120	32.2	37.5	41.66	50.	46.87	56.24	52.08	62.5	57.29	68.75
D	$\frac{3}{8}$ = 0.375		$\frac{7}{16}$ = 0.4375		$\frac{1}{2}$ = 0.5		$\frac{9}{16}$ = 0.5625		$\frac{5}{8}$ = 0.625	
36	208.3	250.	242.	290.4	277.8	333.3	312.4	375.	347.2	416.6
38	197.3	237.	230.3	276.3	263.1	315.8	296.	355.2	329.	394.6
40	187.4	225.	218.7	242.5	250.	300.	281.4	337.8	312.4	374.8
42	178.6	214.3	208.3	249.9	238.	285.6	267.6	321.2	297.4	357.2
44	170.4	204.5	198.8	238.6	227.2	272.7	255.6	306.6	284.	340.8
48	156.2	187.5	182.2	218.6	208.3	250.	234.2	281.2	260.4	312.4
54	138.9	165.7	162.	194.4	185.2	220.2	208.2	250.	231.4	277.8
60	125.	150.	145.7	174.9	166.6	200.	187.4	226.8	208.2	250.
66	113 6	136.3	132.5	159.	151.5	181.8	170.4	204.4	189.4	227.2
72	104.1	125.	121.4	145.7	138.9	166.6	156.2	187.5	173.6	208.2
78	96.15	115.8	112.4	134.9	128.8	153.8	144.5	173.4	160.2	192.3
84	89.28	107.1	104.1	124.9	119.	142.8	133.9	160.7	148.8	178.5
90	83.33	100.	97.21	116.6	111.1	133.3	125.	150.	138.9	166.6
96	78.12	93.74	91.	109.2	104.	124.8	117.1	140.5	130.2	156.2
102	73.53	88.23	85.78	102.9	98.04	117.6	110.3	132.3	122.5	147.
108	69.45	82.85	81.01	97.21	92.6	110.1	104.1	124.9	115.7	138.9
120	62.5	75.	73.86	88 63	83.33	100.	93.74	112.5	104.1	125.

TABLE XXI.

Boiler Plates Stamped 65,000 lbs. Safety-strain $\frac{1}{6}$=10833.3.

Diameter of boiler, inches.	Thickness of boiler-plate in fractions of an inch.									
	$\frac{3}{16}=0.1875$ Riveting.		$\frac{1}{4}=0.25$ Riveting.		$\frac{9}{32}=0.28125$ Riveting.		$\frac{5}{16}=0.3125$ Riveting.		$\frac{11}{32}=0.34375$ Riveting.	
	Single.	Double.	Single.	Double.	Single.	Double.	Single.	Double.	Single.	Double.
D	Pressures.		Pressures.		Pressures.		Pressures.		Pressures.	
36	112.8	135.4	150.4	180.5	169.2	203.	188.	225.6	206.8	248.1
38	106.9	128.3	142.5	171.	160.3	192.4	178.2	213.8	196.	235.2
40	101.5	121.8	135.4	162.5	152.3	182.8	169.3	203.1	186.2	223.4
42	96.72	116.	128.9	154.7	145.	174.	161.2	193.5	177.3	212.8
44	92.32	110.8	123.1	147.7	138.5	166.2	153.9	184.7	169.3	203.1
48	84.63	101.5	112.8	135.4	126.9	152.3	141.	169.3	155.1	186.3
54	75.21	90.25	100.3	120.3	112.8	135.4	125.4	150.4	137.9	165.5
60	67.7	81.24	90.27	108.3	101.5	121.8	112.8	135.4	124.1	148.9
66	61.55	73.86	82.	98.4	92.3	110.7	102.6	123.1	112.8	135.4
72	56.42	67.7	75.22	90.26	84.61	101.5	94.	112.8	103.4	124.1
78	52.	62.4	69.44	83.33	78.12	93.74	86.8	104.1	95.45	114.5
84	48.36	58.	64.48	77.37	72.54	87.05	80.6	96.72	88.66	106.4
90	45.13	54.15	60.18	72.21	67.69	81.23	75.2	90.24	82.72	99.26
96	42.31	50.77	56.37	67.64	63.44	76.13	70.52	84.63	77.57	93.09
102	39.82	47.75	53.1	63.72	59.73	71.68	66.37	79.65	73.01	87.61
108	37.61	45.12	50.15	60.15	56.42	67.71	64.7	75.2	68.95	82.74
120	38.85	40.62	45.13	54.16	50.77	60.93	56.42	67.71	62.06	74.48
D	$\frac{3}{8}=0.375$		$\frac{7}{16}=0.4375$		$\frac{1}{2}=0.5$		$\frac{9}{16}=0.5625$		$\frac{5}{8}=0.625$	
36	225.6	271.	263.2	315.8	300.8	360.9	338.4	406.	376.	451.2
38	213.8	256.6	249.4	299.3	285.1	342.	320.6	384.8	356.4	427.6
40	203.1	243.8	236.9	284.3	270.1	325.	304.6	365.6	338.6	406.2
42	193.5	232.2	225.6	270.7	257.9	309.5	290.	348.	322.4	387.
44	184.7	221.6	215.4	258.5	246.2	295.4	277.	332.4	307.8	369.4
48	169.3	203.1	197.4	236.9	225.7	270.8	253.8	304.6	282.	338.6
54	150.4	180.6	175.5	210.6	200.6	240.7	225.6	270.8	250.8	300.8
60	135.4	162.5	158.	189.5	180.5	216.6	203.	243.6	225.6	270.8
66	123.1	147.7	143.5	172.2	164.	196.8	184.6	221.4	205.2	246.2
72	112.8	135.4	131.6	157.9	150.4	180.5	169.2	203.	188.	225.6
78	104.1	125.	121.4	145.7	138.9	166.6	156.2	187.5	173.6	208.2
84	96.72	116.	112.8	135.4	128.9	154.7	145.1	174.1	161.2	193.4
90	90.24	108.3	105.3	126.4	120.3	144.4	135.4	162.4	150.4	180.5
96	84.63	101.5	98.68	118.4	112.7	135.3	126.9	152.2	141.0	169.2
102	79.65	95.5	92.92	111.6	106.2	127.4	119.4	143.3	132.7	159.3
108	75.2	90.3	87.76	105.3	100.3	120.3	112.8	135.4	125.4	150.4
120	67.71	81.25	83.98	100.8	90.26	108.3	101.5	121.8	112.8	135.4

TABLE XXII.

Boiler Plates Stamped 70,000 lbs. Safety-strain $\frac{1}{6}$ = 11666.6.

Diameter of boiler, inches.	Thickness of boiler-plate in fractions of an inch.									
	$\frac{3}{16} = 0.1875$ Riveting.		$\frac{1}{4} = 0.25$ Riveting.		$\frac{9}{32} = 0.28125$ Riveting.		$\frac{5}{16} = 0.3125$ Riveting.		$\frac{11}{32} = 0.34375$ Riveting.	
	Single.	Double.	Single.	Double.	Single.	Double.	Single.	Double.	Single.	Double.
D	Pressures.		Pressures.		Pressures.		Pressures.		Pressures.	
36	121.5	145.8	164.2	197.1	183.3	220.	202.5	243.	222.7	267.5
38	116.	139.2	153.5	184.2	172.7	217.2	191.9	230.2	211.	253.2
40	109.3	131.2	145.8	174.9	164.	196.8	182.3	218.7	200.5	240.6
42	104.1	125.	138.9	166.6	156.2	187.5	173.6	208.3	190.9	229.1
44	99.42	119.3	132.5	159.	149.1	178.9	165.7	198.8	182.2	218.7
48	91.13	109.3	121.5	145.3	136.7	164.	151.9	182.3	167.1	200.5
54	81.	97.2	108.	129.6	121.5	145.8	135.	162.	148.5	178.2
60	72.9	87.48	97.2	116.6	109.3	131.2	121.5	145.8	133.6	160.4
66	66.3	79.56	88.37	106.	99.43	119.3	110.5	132.5	121.5	145.8
72	60.75	72.9	81.	97.2	91.1	109.3	101.2	121.5	111.3	133.6
78	56.1	67.32	74.7	89.64	80.39	96.47	93.47	112.2	102.8	123.4
84	52.	62.4	69.4	83.28	78.1	93.72	86.8	104.1	95.45	114.5
90	48.6	58.32	64.8	77.77	72.9	87.48	81.	97.2	89.1	106.9
96	45.5	54.6	60.8	72.96	68.37	82.05	75.95	91.14	83.54	101.2
102	42.9	51.3	57.2	68.6	64.35	77.22	71.5	85.8	78.65	94.38
108	40.5	48.6	54.	64.8	60.75	72.9	67.5	81.	74.25	89.1
120	36.45	43.74	48.6	58.32	54.68	65.61	60.76	72.9	66.83	80.2
D	$\frac{3}{8} = 0.375$		$\frac{7}{16} = 0.4375$		$\frac{1}{2} = 0.5$		$\frac{9}{16} = 0.5625$		$\frac{5}{8} = 0.625$	
36	243	291.6	285.7	342.9	328.5	394.2	366.6	440.	405.	486.
38	230.2	276.3	269.5	323.4	307.	368.4	345.4	434.4	383.8	460.4
40	218.7	262.4	255.1	306.1	291.6	349.9	328.	393.6	364.6	437.4
42	208.3	250.	243.	291.6	277.7	333.3	312.4	375.	347.2	416.6
44	198.8	238.	231.9	278.3	265.	318.	298.2	357.8	331.4	397.6
48	182.3	218.7	212.6	255.1	243.	290.6	273.4	328.	303.8	364.6
54	162.	194.4	189.	226.8	216.	259.2	243.	291.6	270.	324.
60	145.8	175.	170.1	204.1	194.4	233.3	218.6	262.4	243.	291.6
66	132.5	159.	154.7	185.6	176.7	212.	198.8	238.6	221.	265.
72	121.5	145.8	141.7	170.1	162.	194.4	182.2	218.6	202.4	243.
78	112.2	134.6	130.8	156.9	149.4	179.3	160.8	192.9	186.9	224.4
84	104.1	125.	121.4	145.7	138.8	166.6	156.2	187.4	173.6	208.2
90	97.2	116.6	113.4	136.1	129.6	155.5	145.8	174.9	162.	194.4
96	91.14	109.3	106.3	127.5	121.6	145.9	136.7	164.1	151.9	182.3
102	85.8	102.6	100.1	120.1	114.4	137.2	128.7	154.4	143.	171.6
108	81.	97.2	94.5	113.4	108.	129.6	121.5	145.8	135.	162.
120	72.9	87.5	85.05	102.	97.2	116.6	109.3	131.2	121.5	145.8

STRENGTH OF BOILER-SHELLS.

§ 74. The steam-pressure per square inch in the boiler, multiplied by the inside diameter of the shell in inches, is the strain on the plates per inch of length of the shell; and as this strain is borne by two sides of the shell, only one-half of it is borne by each side.

S = ultimate strength in pounds per square inch of section of the plate.
t = thickness of the plate in fractions of an inch.
D = inside diameter of the boiler in inches.
p = steam-pressure in pounds per square inch above that of the atmosphere.

§ 75. Ultimate Strength of Solid Shell without Riveted Joints.

Steam-pressure, $p = \frac{2\,t\,S}{D}$. 9

Diameter of boiler, $D = \frac{2\,t\,S}{p}$. 10

Thickness of plate, $t = \frac{D\,p}{2\,S}$. 11

Breaking-strain, $S = \frac{D\,p}{2\,t}$. 12

§ 76. Safety Strength of Solid Shell without Riveted Joints (¼ of the Ultimate Strength).

Steam-pressure, $p = \frac{t\,S}{2\,D}$. 13

Diameter of boiler, $D = \frac{t\,S}{2\,p}$. 14

Thickness of plate, $t = \frac{2\,D\,p}{S}$. 15

Breaking-strain, $S = \frac{2\,D\,p}{t}$. 16

STRENGTH OF SINGLE-RIVETED JOINTS.

§ 77. The post-office engineers pierce the sheets of post-stamps with small holes around each stamp in order to make the sheet tear easily for separating the stamps. This is a practical illustration of the effect of punching holes in the boiler-plates for the riveted joints. The plate is weakened in proportion as the diameter of the rivet is to the distance between the centres of rivets. Suppose the diameter of the rivet to be $d = 1$ and distance between centres $D = 3$, then the strength of the solid plate is to that of the punched plate as

$$1 : \frac{D - d}{D} = 1 : \frac{3 - 1}{3} = 1 : 0.666.$$

That is, the strength of the punched plate is only 66 per cent., or $\frac{2}{3}$ of that of the solid plate.

The static condition of riveted joints is that the sheering strain on the rivet is equal and opposite to the tearing strain on the plate, and the strength to resist these two strains must therefore be alike for the greatest strength of the joint.

It has been found by experiments that the sheering and tearing strength of wrought iron are nearly alike per section strained, and the slight difference varies either way according to the particular iron experimented upon, but on an average the sheering strength appears to have some advantage over that of tearing.

Assuming these two strengths to be alike, the section of the rivet should be equal to the section of the plate between the rivets.

d = diameter of the rivet.

δ = distance between centres of rivets.

t = thickness of plate.

Areas of sections, $0.7854\, d^2 = t\, (\delta - d)$. $\qquad \delta = \frac{d}{t}\, (0.7854\, d + t)$.

The proportion between d and t averages in practice $2\, t = d$—that is, the diameter of the rivet is made twice the thickness of the plate. For thin plates the diameter of the rivet is made larger, and for thick plates smaller, than $d = 2\, t$, as will be seen in the accompanying table, which is set up from practice.

Assuming that $d = 2\, t$ or $t = 0.5\, d$, which, inserted for t in the above formula, will give the proportion between d and δ—namely,

$$0.7854\, d^2 = 0.5\, d\, (\delta - d) \quad \text{and} \quad 0.5824\, d = 0.5\, (\delta - d).$$

Distance $\delta = 2.57\, d$ between centres of rivets.

This is the proportion of δ and d, as used in practice for $\frac{3}{4}$-inch plate, but the diameter of the rivet is then made much less than $2\, t$.

The punching of holes in the boiler-plate disturbs the fibres for some distance around the hole, and thus diminishes the strength, so that the section between the rivets is weaker than an equal section of the same plate not punched. This weakening amounts to from 10 to 20 per cent., according to experiment, with different kinds of iron. Allowing 37 per cent. of section punched away by the hole and 13 per cent. for disturbing the fibres by punching, there remains only 50 per cent. of strength of the solid plate in the single-riveted joint to be relied upon for safety in practice.

Experiments with strength of single-riveted joints have given as high as 70 per cent. of that of the solid plate; but the writer is not disposed to rely upon those experiments in practice of boiler-making, for which reason only 50 per cent. is allowed in the following formulas.

§ 78. Bursting Strength of Single-riveted Joints in Boiler-shells.

Notation of letters is the same as before repeated.

Steam-pressure, $p = \frac{t\,S}{D}$ 17

Diameter of boiler, $D = \frac{t\,S}{p}$ 18

Thickness of plate, $t = \frac{D\,p}{S}$ 19

Breaking-strain, $S = \frac{D\,p}{t}$ 20

The safety strength of materials should not be taken more than 25 per cent. of the ultimate strength.

§ 79. Safety Strength of Single-Riveted Joints with Punched Holes in Boiler Shells.

Steam-pressure, $p = \frac{t\,S}{4\,D}$ 21

Diameter of boiler, $D = \frac{t\,S}{4\,p}$ 22

Thickness of plate, $t = \frac{4\,D\,p}{S}$ 23

Breaking-strain, $S = \frac{4\,D\,p}{t}$ 24

Example. A steam-boiler of $D = 147$ inches diameter is to carry $p = 60$ pounds steam-pressure, and the thickness of plates $t = \frac{5}{8}$ of an inch. Required what stamp the plates must have?

$$S = \frac{4 \times 147 \times 60 \times 8}{5} = 54096.$$

The breaking-strain of the iron plates should be 54096 pounds to the square inch. By the government rule, Formula 4, the stamp need only be 40572.

§ 80. The government rule allows the boilers to be 25 per cent. weaker than by Formulas 21 to 24 inclusive. It is difficult to guard against all carelessness in boiler-making. When the holes in the plates are not punched to properly match one another, they form an eccentric opening, through which a drift is driven to make the holes concentric. This drift does not only overstrain the iron, but inclines the hole so that the rivet will not be at right angles to the plate. The strength of such a rivet may be only 20 per cent. of that of a properly riveted hole. It is almost impracticable to punch the holes in boiler-plates sufficiently correct to match one another, as required for proper work. The strength of single-riveted joints with punched holes should therefore not be taken over 50 per cent. of that of the solid plate.

For drilled holes known to be well fitted, 60 per cent. may be trusted upon for single-riveted joints.

§ 81. **Safety Strength of Single-riveted Joints with Drilled Holes in Boiler Shells.**

Steam-pressure, $p = \frac{0.3\,t\,S}{D}$ 25

Diameter of boiler, $D = \frac{0.3\,t\,S}{p}$ 26

Thickness of plate, $t = \frac{D\,p}{0.3\,S}$ 27

Breaking-strain, $S = \frac{D\,p}{0.3\,t}$ 28

§ 82. It is impracticable to proportion the riveted joints so perfectly that the shearing strength of the rivet be equal to the tearing strength of the plate, for the actual strength of the iron varies more than does the proportion of dimensions of the joint.

The following table gives the proportions of single-riveted joints to the nearest 16th of an inch as used in practice.

It will be seen in the table that the section of the plate between the rivets is greater than the section of the rivet, except for one-eighth of an inch plate.

For drilled holes make the distance between the centres of the rivets one-eighth ($\frac{1}{8}$) of an inch less than that for punched holes.

TABLE XXIII.

Proportion of Single-riveted Lap-joints with Punched Holes.

Thickness of plate.	Rivets. Diameter.	Rivets. Length.	Distance betw. cent.	Lap of joint.	Area of rivet.	Area of plate.	Per cent. of solid
t	d	l	δ	inches.	sq. inch.	sq. inch.	plate.
1/8	5/16	1/2	7/8	1.1/4	0.0767	0.07031	64
3/16	7/16	3/4	1.5/16	1.1/2	0.1503	0.16406	66
1/4	1/2	1.1/8	1.1/2	1.3/4	0.1963	0.25000	66
5/16	5/8	1.3/8	1.7/8	2 in.	0.3067	0.39062	66
3/8	3/4	1.11/16	2.1/4	2.1/4	0.4417	0.56250	66
7/16	13/16	1.15/16	2.3/8	2.3/8	0.5184	0.68359	65
1/2	7/8	2.1/4	2.1/2	2.1/2	0.6013	0.75250	64
9/16	1 in.	2.1/2	2.5/8	2.5/8	0.7854	0.91406	63
5/8	1.1/16	2.13/16	2.3/4	2.7/8	0.8904	1.05468	62
11/16	1.1/8	3.1/8	2.7/8	3.1/8	0.9940	1.03125	61
3/4	1.3/16	3.5/8	3 in.	3.3/8	1.3603	1.35937	60
13/16	1.5/16	3.11/16	3.1/4	3.5/8	1.3605	1.57422	60
7/8	1.3/8	3.15/16	3.1/2	4 in.	1.4840	1.85937	60
15/16	1.1/2	4.1/4	3.3/4	4.1/4	1.767	2.10937	60
1 in.	1.5/8	4.1/2	4 in.	4.5/8	2.073	2.375	60

DOUBLE-RIVETED LAP-JOINTS.

§ 83. Double-riveted joints, if properly proportioned, increase the strength of the boiler about 40 per cent. on account of the rivets being spaced farther apart, leaving more section of plate between them to resist the strain. The rivets are arranged in two rows, zig-zag, over one another, as shown in the accompanying illustration. For the greatest strength the distance between the rivets in the direction of the joint should be double the distance between the centre lines of the two rows, and the rivets will then form a right angle, or 90°, with one another.

The distance between the rivets in the direction of the joint can be made 42 to 50 per cent. greater than between rivets in single-riveted joints.

The diagonal distance between centres of rivet should be made equal to the distance in the direction of the joints in single riveting.

Fig. 4.

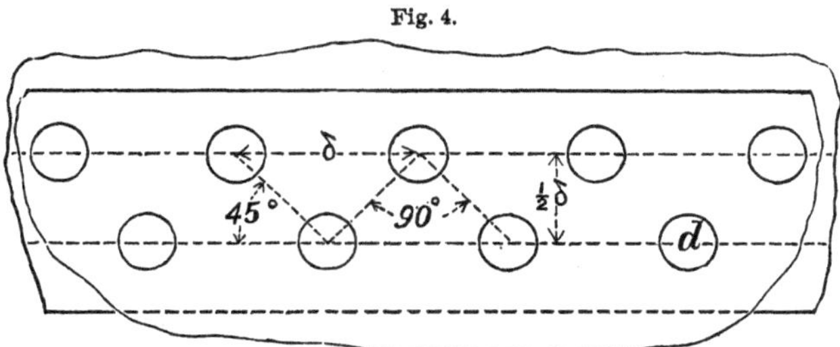

Double-riveted joints with punched holes, proportioned according to this rule, should be 40 per cent. stronger than single-riveted joints, and with drilled holes about 60 per cent. stronger.

§ 84. Safety Strength of Double-riveted Lap-joints with Punched Holes in Boiler-shells.

Steam-pressure, $p = \dfrac{0.35\,t\,S}{D}$. 29

Diameter of boiler, $D = \dfrac{0.35\,t\,S}{p}$. 30

Thickness of plate, $t = \dfrac{D\,p}{0.35\,S}$. 31

Breaking-strain, $S = \dfrac{D\,p}{0.35\,t}$. 32

In the following tables for double-riveted lap-joints, one is headed A for drilled holes and the other B for punched holes, their difference being only in the distance of rivets. When the boiler-plates are stamped a low figure, say 45000, and the rivets are known to be of extra good quality, then table B should be used for drilled holes.

For boiler-iron of high stamp, say 65000, and the rivets of ordinary quality, then table A should be used for punched holes. The dimensions in the tables are given to the nearest 16ths of an inch.

TABLE XXIV.

A. Proportions of Double-riveted Lap-joints with Drilled Holes.

Thickness of plate.	Rivets.		Distance between Rivets.		Dist. between Cent. lines.	Lap of joint.
	Diameter.	Length.	Central.	Diagonal.		
t	d	l	δ			
1/8	5/16	1/2	1.1/4	7/8	5/8	1.5/8
3/16	7/16	3/4	1.7/8	1.5/16	15/16	2.3/16
1/4	1/2	1.1/8	2.1/8	1.1/2	1.1/16	2.9/16
5/16	5/8	1.3/8	2.5/8	1.7/8	1.5/16	3.1/4
3/8	3/4	1.11/16	3.3/16	2.1/4	1.3/8	3.7/16
7/16	13/16	1.15/16	3.3/8	2.3/8	1.11/16	4 inches.
1/2	7/8	2.1/4	3.9/16	2.1/2	1.13/16	4.1/4
9/16	1 inch.	2.1/2	3.3/4	2.5/8	1.7/8	4.1/2
5/8	1.1/16	2.13/16	3.7/8	2.3/4	1.15/16	4.7/16
11/16	1.1/8	3.1/8	4.1/16	2.7/8	2.1/16	5.1/8
3/4	1.3/16	3.5/8	4.1/4	3 inches.	2.1/8	5.7/16
13/16	1.5/16	3.11/16	4.9/16	3.1/4	2.5/16	5.7/8
7/8	1.3/8	3.15/16	4.15/16	3.1/2	2.1/2	6.7/16
15/16	1.1/2	4.1/4	5.5/16	3.3/4	2.11/16	6.15/16
1 inch.	1.5/8	4.1/2	5.5/8	4 inches.	2.7/8	7.1/2

TABLE XXV.

B. Proportion of Double-riveted Lap-joints with Punched Holes.

Thickness of plate.	Rivets.		Distance between Rivets.		Dist. between Cent. lines.	Lap of joint.
	Diameter.	Length.	Central.	Diagonal.		
t	d	l	δ			
1/8	5/16	1/2	1.3/8	1 inch.	11/16	1.7/8
3/16	7/16	3/4	2 inches.	1.7/16	1 inch.	2.1/8
1/4	1/2	1.1/8	2.1/4	1.9/16	1.1/8	2.3/8
5/16	5/8	1.3/8	2.13/16	2 inches.	1.7/16	2.3/4
3/8	3/4	1.11/16	3.3/8	2.3/8	1.11/16	3.3/8
7/16	13/16	1.15/16	3.9/16	2.1/2	1.13/16	3.1/4
1/2	7/8	2.1/4	3.13/16	2.11/16	1.15/16	3.3/4
9/16	1 inch.	2.1/2	4 inches.	2.13/16	2 inches.	4.1/4
5/8	1.1/16	2.13/16	4.1/8	2.15/16	2.1/16	4.3/4
11/16	1.1/8	3.1/8	4.5/16	3.1/16	2.3/16	5.1/8
3/4	1.3/16	3.5/8	4.1/2	3.3/16	2.1/4	5.3/8
13/16	1.5/16	3.11/16	4.7/8	3.7/16	2.7/16	5.5/8
7/8	1.3/8	3.15/16	5.1/4	3.11/16	2.5/8	6.1/8
15/16	1.1/2	4.1/4	5.5/8	3.15/16	2.9/16	6.5/8
1 inch.	1.5/8	4.1/2	6 inches.	4.3/16	3 inches.	7 inches.

§ 85. **Safety Strength of Double-riveted Lap-joints with Drilled Holes in Boiler-shells.**

Steam-pressure, $p = \frac{0.4\,t\,S}{D}$ 33

Diameter of boiler, $D = \frac{0.4\,t\,S}{p}$ 34

Thickness of plate, $t = \frac{D\,p}{0.4\,S}$ 35

Breaking-strain, $S = \frac{D\,p}{0.4\,t}$ 36

Example 33. What pressure can be carried with safety in a boiler of $D = 72$ inches diameter, made of steel plates stamped $S = 75000$ pounds tensile strength and $t = \frac{1}{2}$ inch thick, when the boiler is double-riveted with drilled holes?

$$p = \frac{0.4 \times 0.5 \times 75000}{72} = 208 \text{ pounds to the square inch.}$$

TABLE XXVI.

§ 86. **Coefficients X for Safety Strength of Lap-joints.**

Construction of Shell.	X	Per cent. of strength.
Solid plate without joints......................	0.5	100
Double-riveted drilled holes................	0.4	80
Double-riveted punched holes..............	0.35	70
Single-riveted drilled holes..................	0.3	60
Single-riveted punched holes..............	0.25	50

Steam-pressure, $p = \frac{X\,t\,S}{D}$ 37

Diameter of boiler, $D = \frac{X\,t\,S}{p}$ 38

Thickness of plate, $t = \frac{D\,p}{X\,S}$ 39

Breaking-strain, $S = \frac{D\,p}{X\,t}$ 40

§ 87. The greatest strain in a cylindrical boiler-shell is in the direction of the circumference, for which the double-riveted joints are first required in the direction of the length of the boiler.

Longitudinal strain, $=\pi D t S = p\frac{\pi}{4}D^2$ 41

Required thickness of metal, $t = \frac{p D}{2 S}$ 42

Transverse strain, $= t S = p D$ . . . 43

Required thickness of metal, $t = \frac{p D}{S}$ 44

That is to say, the longitudinal strain is only one-half of the transverse strain, or that single-riveted joints with punched holes around the boiler are stronger than double-riveted joints with drilled holes longitudinally.

Double-riveted joints are therefore required only longitudinally.

STRENGTH OF FLUES AND TUBES FOR EXTERNAL PRESSURE TO COLLAPSE.

§ 88. The most reliable experiments on this subject yet made are those of the late Mr. Fairbairn, who stated that the strength of the flue is inversely as its length, but he proposed different coefficients for different lengths.

By analyzing closely the results of Mr. Fairbairn's experiments and by using constant coefficients, we find that the strength is inversely as the square root of the length of the flue or tube.

The following formulas are deduced from the results of those experiments without regard to the formulas proposed by Mr. Fairbairn.

D = diameter of the flue or tube in inches.
L = length of the same in feet.
t = thickness in fractions of an inch of the iron in the flue.
p = steam pressure in pounds per square inch.
S = tensile strength per square inch of iron in the flues.

§ 89. Collapsing Strength of Flues subjected to External Pressure.

Steam-pressure, $p = \dfrac{4\,S\,t^2}{D\sqrt{L}}$. 41

Diameter of flue, $D = \dfrac{4\,S\,t^2}{p\sqrt{L}}$. 42

Thickness of metal, $t = \sqrt{\dfrac{p\,D\sqrt{L}}{4\,S}}$. . . . 43

Length of flue, $L = \left(\dfrac{4\,S\,t^2}{p\,D}\right)^2$. 44

Assuming one-fourth of the collapsing strength as safety for the flue, the formulas will simply dispense with the coefficient 4.

§ 90. Safety Strength of Flues and Tubes from Collapsing by External Pressure.

Steam-pressure, $p = \dfrac{S\,t^2}{D\sqrt{L}}$. 45

Diameter of flue, $D = \dfrac{S\,t^2}{p\sqrt{L}}$. 46

Thickness of iron, $t = \sqrt{\dfrac{p\,D\sqrt{L}}{S}}$. . . . 47

Length of flue, $L = \left(\dfrac{S\,t^2}{p\,D}\right)^2$ 48

Example 45. A flue made of iron $S = 50000$ pounds strength is $D = 18$ inches in diameter and $L = 16$ feet long, by $t = \frac{3}{8}$ of an inch metal. Required what steam-pressure the flue can stand with safety?

$$p = \frac{50000 \times 3^2}{18 \times \sqrt{16} \times 8^2} = 97.66 \text{ pounds to the square inch.}$$

STAYING OF FLAT BOILER SURFACES.

§ 91. Flat surfaces subject to steam-pressure in boilers must be stayed in order to keep their proper flat position as intended, and thus the whole steam-pressure on such surface must be borne by stays.

A = area in square inches to be stayed. a = section area of each stay in square inches. n = number of stays required. p = steam-pressure in pounds per square inch. S = tensile strength of the iron in the stays. D = distance between the stays in inches.

$A\,p = a\,n\,S$ and $a = \dfrac{A\,p}{n\,S}$. 44 | Pressure on each stay, $a\,S = \dfrac{A\,p}{n} = D^2 p$. 46

Number of stays, $n = \dfrac{A\,p}{a\,S}$. 45 | Distance, $D = \sqrt{\dfrac{a\,S}{p}}$. 47

Suppose the stays to be round of diameter d; then $a = \dfrac{\pi}{4} d^2$.

$$D = d\sqrt{\frac{\pi\,S}{4\,p}} = 0.886d\sqrt{\frac{S}{p}}. \quad . \quad . \quad . \qquad 48$$

Allowing 28 per cent. for safety of the ultimate strength of stays, we have

Safety Formulas for Stay-bolts.

Diameter of stay, $d = 4\,D\sqrt{\dfrac{p}{S}}$. 49 | Steam-pressure, $p = \dfrac{d^2\,S}{16D^2}$. . 51

Distance apart, $D = \dfrac{d}{4}\sqrt{\dfrac{S}{p}}$. . 50 | Iron required, $S = \dfrac{16\,D^2\,p}{d^2}$. 52

Example 50. The iron for stay-bolts in a steam-boiler is $d = 1$ inch diameter and $S = 62500$ pounds strength, to be used in a pressure of $p = 64$ pounds to the square inch. Required the distance apart of the stays?

$$D = \frac{1}{4}\sqrt{\frac{62500}{64}} = 8 \text{ inches.}$$

The strength of all the connections of the stays must be equal to that of the solid stay. When the sections of the stays are square or rectangular, the area must be equal to that corresponding to the diameter d of the round iron.

The following table is calculated for stays of one inch diameter; but when the stays are more or less, the spaces between them should be that much more or less; for instance, if the stays are $\frac{3}{4}$ inch diameter, the spaces in the table should be multiplied by $\frac{3}{4}$, and so on.

TABLE XXVII.

Distance in Inches between Boiler-stays One Inch in Diameter.

Steam pressure.	Breaking strain in pounds per square inch of stay. 45,000.	50,000.	55,000.	60,000.	65,000.	70,000.
p.	*Inches.*	*Inches.*	*Inches.*	*Inches.*	*Inches.*	*Inches.*
25	10.6	11.2	11.7	12.5	12.7	13.2
30	9.68	10.2	10.7	11.4	11.6	12.
35	8.96	9.45	9.9	10.5	10.8	11.1
40	8.38	8.84	9.26	9.84	10.1	10.4
45	7.9	8.34	8.74	9.28	9.51	9.84
50	7.5	7.9	8.28	8.8	9.02	9.34
55	7.15	7.54	7.9	8.4	8.6	8.9
60	6.85	7.22	7.56	8.04	8.24	8.52
65	6.58	6.94	7.26	7.72	7.91	8.18
70	6.34	6.68	6.99	7.43	7.62	7.88
75	6.12	6.45	6.75	7.18	7.36	7.61
80	5.93	6.25	6.54	6.96	7.12	7.38
85	5.75	6.07	6.35	6.75	6.91	7.15
90	5.59	5.89	6.17	6.56	6.72	6.96
95	5.43	5.73	6.	6.39	6.54	6.77
100	5.3	5.6	5.86	6.23	6.37	6.6
110	5.05	5.32	5.58	5.93	6.08	6.29
120	4.84	5.1	5.35	5.68	5.82	6.02
130	4.56	4.9	5.13	5.46	5.58	5.79
140	4.48	4.73	4.95	5.26	5.38	5.58
150	4.33	4.56	4.78	5.08	5.2	5.39
160	4.19	4.42	4.62	4.92	5.03	5.21
170	4.06	4.29	4.49	4.78	4.88	5.06
180	3.95	4.17	4.36	4.64	4.75	4.91
190	3.85	4.06	4.25	4.52	4.63	4.79
200	3.74	3.95	4.14	4.4	4.51	4.66
210	3.66	3.86	4.04	4.3	4.4	4.56
220	3.57	3.77	3.94	4.2	4.3	4.44
230	3.5	3.68	3.86	4.1	4.2	4.35
240	3.42	3.61	3.78	4.02	4.11	4.26
250	3.35	3.53	3.7	3.93	4.03	4.17
260	3.29	3.47	3.63	3.86	3.95	4.1
270	3.23	3.4	3.56	3.79	3.88	4.02
280	3.16	3.34	3.5	3.71	3.8	3.94
290	3.11	3.28	3.43	3.65	3.74	3.87
300	3.06	3.23	3.38	3.6	3.68	3.81

STEAM-POWER WITHOUT FIRE.

Fig. 5.

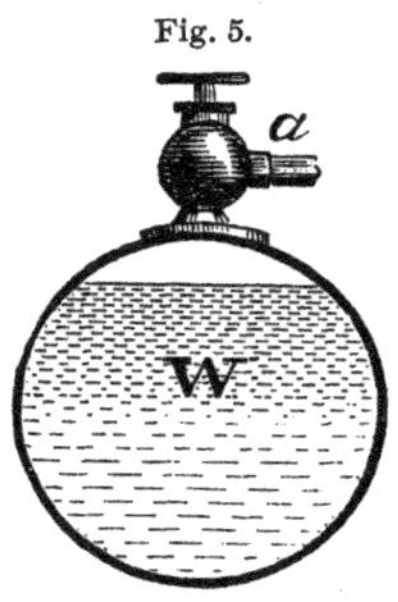

§ 92. When water is heated under high-pressure in a closed vessel, the work so stored can be utilized for motive-power after the fire is withdrawn.

Fig. 5 represents a section of a cylindrical vessel nearly full of hot water, above which surface steam is to be conducted to a motor through the valve and pipe *a*.

Suppose no heat to radiate from the vessel and no discharge of steam, there will then only be a static pressure corresponding to the temperature of the water, and no work is performed.

The combination of heat, water and steam enclosed in a vessel constantly tends to keep the presence and temperature in equilibrium—that is, a given pressure corresponds with a certain temperature. Therefore, if steam is allowed to escape through the pipe *a*, the temperature and pressure in the steam-room will be lowered below that in the water, the result of which is that the excess of temperature in the water will generate more steam to establish equilibrium.

W = pounds of water in the vessel.

T = temperature Fahr. of the steam and water.

P = steam-pressure in pounds per square inch above vacuum in the vessel.

C = cubic feet of steam used per double stroke in a steam-engine.

n = double strokes per minute of the steam piston.

p = steam-pressure in pounds per square inch above that of the atmosphere in the cylinder.

H = units of heat per pound in the water before the engine is started.

H' = units of heat per pound of the water in the vessel after the engine has made n revolutions.

h = units of heat per cubic foot of the steam driving the engine.

w = pounds of water passed through the engine in form of steam.

$\mathfrak{P}$ = weight per cubic foot of steam.

§ 93. The primitive number of units of heat in the vessel is WH, and after the engine has made n revolutions, that heat will be reduced to

$$H'(W-w) = WH - Cnh. \quad . \quad . \quad . \quad 1$$

The heat consumed by the engine will then be Cnh.

The weight w of steam passed through the engine is $w = C \mathfrak{P} n$, which, inserted for w in Formula 1, gives

$$H'(W - C \mathfrak{P} n) = W H - C n h. \quad . \quad . \quad 2$$

Revolutions, $$n = \frac{W(H - H')}{C(h - \mathfrak{P} H')}. \quad . \quad . \quad . \quad . \quad 3$$

Example 3. A vessel containing 200 cubic feet of water of temperature $T = 358°$, corresponding to a pressure of $P = 150$ pounds to the square inch, supplies steam which is wire-drawn to a pressure of $p = 30$ pounds to an engine using $C = 1.5$ cubic feet of steam for each revolution.

Required how many revolutions the engine will make before the steam-pressure in the vessel is reduced to $p = 30$ or $P = 45$ pounds?

The weight of water in the boiler is

$$W = 200 \times 56.073 = 11214.6 \text{ pounds.}$$

$$H = 330.75. \quad H' = 241.32. \quad \mathfrak{P} = 0.11111. \quad h = 129.51.$$

See tables Nystrom's *Pocket-Book* for these data.

Revolutions, $$n = \frac{11214.6(330.75 - 241.32)}{1.5(129.51 - 0.1111 \times 241.32)} = 6570.6.$$

The water, evaporated to steam, will be

$$w = 1.5 \times 0.11111 \times 6570.6 = 1095.1 \text{ pounds,}$$

or nearly 10 per cent. of the primitive water in the vessel.

Assuming the engine to make 80 revolutions per minute, it will run

$$\frac{6570.6}{80 \times 60} = 1.369 \text{ hours, with the steam generated in the vessel.}$$

Practically, the radiation of heat from the vessel and steam-pipe will reduce this time perhaps 15 cents.

Dr. Emile Lamm of New Orleans constructed a locomotive upon the above principle with heated water without fire, and which was used on General Beauregard's road in the year 1872.

PERMANENT GASES.

§ 94. Permanent gases, in distinction from vapors, are those that cannot be condensed to liquid under ordinary temperatures and pressures.

Oxygen, *nitrogen* and *hydrogen* are the principal permanent gases, and any mechanical mixture of either two or all the three will remain a permanent gas like atmospheric air, which is a mixture of oxygen and nitrogen; but any chemical combination of either two or all the three becomes a vapor which is condensable to liquid like that of oxygen and hydrogen, forming steam, which condenses to water under temperature 212° Fahr. and freezes solid at 32°.

ELASTICITY OF PERMANENT GASES.

§ 95. Permanent gases are perfectly elastic—that is, the product of volume and pressure of a definite weight of gas will remain constant under constant temperature. For instance, if the volume is compressed to one-half, the pressure will be double; and if again expanded to its primitive volume, the original pressure will be restored if the temperature remains constant. When the temperature varies, the product of volume and pressure will also vary in a direct ratio to the difference of temperature.

Call V and P volume and pressure of a definite weight of gas of temperature T. v and p = volume and pressure of the same gas, but of temperature t. P and p mean the actual pressures of the gas above vacuum.

Then $$\frac{V P}{v p} = 1 + \frac{T - t}{c} \quad . \quad . \quad . \quad . \quad 1$$

That is to say, the ratio of the products of volume and pressure increases arithmetically as the difference of temperature.

The experiments on elasticity of permanent gases made by Regnault and Rudberg show that c is constant for any difference of temperature within the limit of those experiments.

Call $v p = 1$ when $t = 32°$, and find the value of $V P$ when $T = 212°$ or a difference in temperature of 180°. Under this condition the experiments of Regnault and Rudberg show that

$$\frac{\mathfrak{V}\,P}{\mathfrak{v}\,p} = 1.365, \text{ that is, } 1+0.365. \quad . \quad . \qquad 2$$

Consequently, $$0.365 = \frac{T-t}{c} = \frac{180^\circ}{c}, \quad . \quad . \quad . \quad . \quad . \qquad 3$$

of which $$c = \frac{180^\circ}{0.365} = 493.15.$$

Then $$\frac{\mathfrak{V}\,P}{\mathfrak{v}\,p} = 1 + \frac{T-t}{493.15} \text{ for all permanent gases.} \qquad 4$$

Drop the fraction 0.15, and say 493.

Assume the pressure to be constant—

That is, $$\frac{P}{p} = 1. \quad . \quad . \quad . \quad . \quad . \quad . \qquad 5$$

Then $$\frac{\mathfrak{V}}{\mathfrak{v}} = 1 + \frac{T^\circ - t}{493}. \quad . \quad . \quad . \quad . \qquad 6$$

and $$\mathfrak{V} = \mathfrak{v}\left(1 + \frac{T^\circ t}{493}\right). \quad . \quad . \quad . \qquad 7$$

Call $\mathfrak{v} = 1$ at the temperature $t = 32^\circ$. Then the volume $\mathfrak{V}$ can be determined by Formula 7 for any other temperature T, and under constant pressure. For instance, suppose the temperature of the volume $\mathfrak{v}$ to be reduced to $T^\circ = -461^\circ$, then

$$\text{the volume } \mathfrak{V} = \mathfrak{v}\left(1 + \frac{-461-32}{493}\right) = \mathfrak{v}\,(1-1) = 0.$$

This implies not only that the volume of a permanent gas can be reduced to nothing, and even negative, but that matter which exists in the universe may be rendered extinct or less than nothing, which is simply preposterous. Therefore c cannot be a constant quantity.

It is generally supposed by scientific men of our days that the temperature 461° below Fahrenheit's zero is an absolute zero or lowest limit of temperature, which hypothesis is based upon the assumption that for all permanent gases

$$\frac{P\,\mathfrak{V}}{p\,\mathfrak{v}} = 1 \times \frac{T-t}{493}.$$

This formula implies that the intervals between the temperatures progress in the same ratio as do the intervals between $P \mathfrak{V} : p \mathfrak{v}$, which the author inclines to doubt.

§ 96. We have yet no experimental data and not sufficient knowledge on the subject by which to contradict the existence of this absolute zero at that place. It is evident, however, that matter cannot be rendered extinct, but that there must exist some low temperature at which the force of expansion of the heat is equal to or less than the force of attraction between the atoms composing the gas, which must then be a liquid, solid or powder of a definite volume; and it is reasonable to suppose that the temperature of that volume can be further reduced.

Considering that water is practically incompressible, we may assume that the atoms of oxygen and hydrogen are there in close contact, and represent the volume of these gases in a liquid or solid state.

One cubic foot of water at 32° weighs 62.4 pounds, of which there are—

54.6 pounds of liquid oxygen in $\frac{1}{3}$ cubic foot.
7.8 pounds of liquid hydrogen in $\frac{2}{3}$ " "
1 pound liquid oxygen = 0.006105 cubic foot.
1 pound liquid hydrogen = 0.08547 " "
1 pound oxygen gas at 32° = 11.28 " "
1 pound hydrogen gas at 32° = 180 " "

1 volume liquid oxygen $= \frac{11.28}{0.006105} = 1847.7$ volumes of oxygen gas at 32°.

1 volume liquid hydrogen $= \frac{180}{0.08547} = 2106$ volumes of hydrogen gas at 32°.

1 volume oxygen gas = 0.0005412 volumes of liquid oxygen.
1 volume of hydrogen gas = 0.0004748 volumes of liquid hydrogen.

Allowing for contraction of the liquid volume by cooling from 32° to −461° or $\mathfrak{T} - \mathfrak{t} = 493°$, at the same rate as ice contracts, about 0.8547 of that at 32°.

Volume of liquid oxygen at −461° is then

$$0.0005412 \times 0.8547 \mathfrak{V} = 0.00046256 \mathfrak{V}.$$

Volume of liquid hydrogen at −461° is

$$0.0004748 \times 0.8547 \mathfrak{V} = 0.00040581 \mathfrak{V}.$$

This should be the ultimate volumes to which gases of oxygen and hydrogen can be reduced by cooling from +32° to −461°.

The oxygen and hydrogen of one cubic foot of water, dissolved into their respective gases, would occupy 1919.9 cubic feet at 32° Fahr., or 2610.66 cubic feet at 212°, and under atmospheric pressure.

§ 97. It is supposed in the preceding calculation that if one cubic foot of water is resolved into its elements and still remain in liquid form, the hydrogen would occupy $\frac{2}{3}$ and the oxygen $\frac{1}{3}$ of the cubic foot; but such would, however, not be the case. The hydrogen would occupy the whole cubic foot, whether the oxygen is in it or not. The atoms of hydrogen may be represented by large potatoes filling a bushel, but the real capacity of the potatoes is only $\frac{2}{3}$ of that bushel; the other $\frac{1}{3}$ can be filled up with buckshot, representing the atoms of oxygen. The potatoes would occupy the same space whether the shot are there or not. Such is the case with hydrogen and oxygen in water; but when these elements are resolved into their respective gases, they will occupy 50 per cent. more volume than when chemically combined in the form of vapor. The result of the preceding calculation is, however, correct.

It is reasonable to suppose that the so-called permanent gases become vapors and finally condense to liquids and freeze to solids at a low temperature, which we have not yet been able to produce, and that there is therefore a limit beyond which the volume of those gases cannot be reduced. The pressure, on the other hand, is reduced to nothing at a low temperature when the vapors condense to liquid and freeze to ice; but that is no proof of an absolute zero having been reached beyond which there exists no temperature.

Steam highly superheated behaves very much like permanent gases; and if experimented upon without knowing the lower temperatures at which it condenses to water and freezes to ice, the inference might be that there exists an absolute zero at which the pressure and volume of steam become nothing, and beyond which there exists no temperature.

Carbonic acid gas under ordinary pressures and temperatures behaves like permanent gases; but at low temperatures and high pressures it becomes a vapor which can be condensed to liquid and even frozen solid.

Water and ice evaporate under low temperatures, as shown by the experiments of Regnault and Dalton. A wet cloth exposed to very cold weather freezes stiff, but finally the ice in it evaporates and leaves the cloth dry.

The formulas which the writer has deduced from the experiments of Regnault and Dalton, indicate that the pressure of aqueous vapor is reduced to nothing at the temperature $-101°$ below Fahr. zero.

Such is most likely the case with all permanent gases—namely, that at some low temperature different for each kind of gas the pressure is reduced to nothing, whilst the volume remains definite, whether in the form of gas, vapor, liquid or solid. Therefore, when the matter is in the form of a gas or vapor at the low temperature where the pressure is reduced to nothing, the force of attraction between its atoms is equal to the force of expansion by heat, and the gas occupies a definite volume like a cloud in the air. Thus, the top of our atmosphere would maintain a smooth surface like the ocean, omitting the disturbance caused by change of temperature and currents of wind below.

§ 98. Within the limit of our practice we can safely use the formula

$$\frac{P\not{V}}{p\mathfrak{V}} = 1 + \frac{\mathfrak{T} - \mathfrak{t}}{493}.$$

Under constant pressure the increase of volume of any permanent gas, per degree of increased temperature—that is, when $\mathfrak{T} - t = 1$ will be $\frac{1}{493} = 0.0020284$.

For simplicity in elucidating the subject and for the formation of tables, it is best to assume a standard temperature, $\mathfrak{t} = 32°$ Fahr., at which all other quantities are compared.

Call $$x = 1 + \frac{\mathfrak{T} - 32}{493} = \frac{P\not{V}}{p\mathfrak{V}}.$$

The value of x is calculated for every degree of temperature from 0° to 500°, for every 10° from 500° to 1200°, and for every 100° from 1200° to 2300°, in Table XXX.

§ 99. Variable Volume under Constant Pressure.

Temperature, $x = \frac{\not{V}}{\mathfrak{V}}$. 1

Heated volume, $\not{V} = \mathfrak{V}x$. 2

Cold volume, $\mathfrak{V} = \frac{\not{V}}{x}$. 3

Example 1. A volume $\mathfrak{V} = 36$ cubic feet of air is to be heated from 32° until the volume is expanded to $\not{V} = 48$ cubic feet. Required the temperature of the expanded volume?

$$x = \frac{48}{32} = 1.5.$$

Find 1.5 in column x in the table, which corresponds to the required temperature, $T = 279$ Fahr.

If the volume V had been heated from a higher temperature, say $t = 60°$, then $60 - 32 = 28$ and $279 + 28 = 307°$, the required temperature.

Example 2. A volume of air $V = 24$ cubic feet is heated from $t = 48°$ to $T = 450°$. Required the volume $\mathcal{V}$?

In this case $48 - 32 = 16$ and $450 + 16 = 466°$. Find x for $466°$, which in the table corresponds to $x = 1.88$.

$$\text{Volume } \mathcal{V} = 24 \times 1.88 = 45.12 \text{ cubic feet.}$$

Example 3. A volume of air $\mathcal{V} = 148$ cubic feet, and of temperature $T = 250°$, is to be cooled down to $t = 32°$. What will be the volume of the cooled air?

Cold volume, $$V = \frac{148}{1.442} = 102.63 \text{ cubic feet.}$$

§ 100. **Variable Pressure under Constant Volume.**

Temperature, $x = \frac{P}{p}$. 4

High pressure, $P = px$. 5

Low pressure, $p = \frac{P}{x}$. 6

Example 4. A volume of permanent gas enclosed in a vessel exerts a pressure of $p = 15$ pounds to the square inch, and is $t = 32°$ in temperature. To what temperature must that gas be elevated in order to increase the pressure to $P = 25$ pounds to the square inch?

$$x = \frac{25}{15} = 1.6666.$$

The required temperature is $T = 361°$.

Had the primitive temperature in the vessel been more or less than 32°, the required temperature would have been that much more or less.

Example 5. A gas of temperature $t = 21°$, enclosed in a vessel under a pressure of $p = 12$ pounds to the square inch, is to be heated to a temperature $T = 180°$. Required the pressure of the heated gas?

In this case $T = 180 + 11 = 191°$.

Pressure $P = 12 \times 1.3224 = 15.8888$ pounds per square inch.

Example 6. The temperature of a permanent gas enclosed in a vessel is $T = 120°$, and pressure $P = 20$ pounds to the square inch, is to be reduced to $t = 5°$. Required the pressure p of the cold gas?

In this case $T = 120 + 5 + 32 = 157$, and $x = 1.2535$.

Pressure, $$p = \frac{20}{1.2535} = 15.95 \text{ pounds per square inch.}$$

§ 101. VOLUME AND PRESSURE BOTH VARIABLE.

Temperature, $$x = \frac{P V}{p v} \quad \ldots \ldots \quad 7$$

High pressure, $$P = \frac{p v x}{V} \quad \ldots \ldots \quad 8$$

Low pressure, $$p = \frac{P V}{v x} \quad \ldots \ldots \quad 9$$

Warm volume, $$V = \frac{p v x}{P} \quad \ldots \ldots \quad 10$$

Cold volume, $$v = \frac{P V}{p x} \quad \ldots \ldots \quad 11$$

Example 7. A volume of air $v = 16$ cubic feet, pressure $p = 15$ pounds to the square inch and temperature 32°, is to be heated until the volume becomes $V = 24$ cubic feet and pressure $P = 20$ pounds to the square inch. Required the temperature of the heated air.

$$x = \frac{20 \times 24}{16 \times 15} = 2.$$

The required temperature is $T = 530°$.

Example 8. A volume of air $V = 42$ cubic feet and temperature $T = 480°$ has been expanded from $v = 28$ cubic feet of temperature $t = 62°$ and pressure $p = 15$ pounds. Required the pressure of the expanded volume?

$62 - 32 = 30°$, and $480 - 30 = 450$. $x = 1.8477$.

Pressure, $$P = \frac{15 \times 28 \times 1.8477}{42} = 18.477 \text{ pounds.}$$

Example 9. The temperature of a permanent gas is $T = 248°$, pressure $P = 48$ pounds and volume $V = 96$ cubic feet. The volume is to be reduced to $v = 72$ cubic feet of temperature $t = 72°$. Required the pressure p?

$$72 - 32 = 40°. \quad 248° - 40° = 208°. \quad x = 1.3569.$$

Pressure, $$p = \frac{48 \times 96}{72 \times 1.3569} = 47 \text{ pounds.}$$

SPECIFIC HEAT OF PERMANENT GASES.

§ 102. The specific heat of a gas is that fraction of a unit of heat required to elevate the temperature of one pound of that gas one degree Fahrenheit. It is constant under constant pressure, but under variable pressure the specific heat is inversely as the square root of the pressure.

TABLE XXVIII.

Specific Heat under Constant Pressure and Temperature 32°.

Kinds of gases.	Pounds per cubic foot.	Cubic foot per pound.	Specific gravity. Water = 1.	Specific gravity. Air = 1.	Specific heat.
	$\mathfrak{P}$	$\mathfrak{C}$			S
Atmospheric air............	0.08042	12.433	0.00130	1.000	0.25
Oxygen gas..................	0.08888	11.251	0.00143	1.104	0.23
Nitrogen gas.................	0.07837	12.760	0.00126	0.972	0.275
Hydrogen gas................	0.00559	178.84	0.00009	0.069	3.3
Carbonic oxide..............	0.07837	12.760	0.00126	0.972	0.288
Carbonic acid................	0.12333	8.108	0.00197	1.527	0.221
Steam..........................	0.05021	19.915	0.00634	0.488	0.475

S = specific heat under constant pressure, as in the table above.

s = mean specific heat under any pressure and volume from 32° to T.

p = 14.7 pounds to the square inch pressure of the gas at $t = 32°$ Fahr.

P = pressure of the same gas at the temperature T.

v = volume in cubic feet of the gas at 32°.

V = volume of the same gas, but of pressure P and temperature T.

W = weight in pounds of the gas experimented upon.

$\mathfrak{P}$ = weight in a fraction of a pound per cubic foot of the gas.

h = units of heat in W pounds of gas elevated from 32° to T, or from a pressure of 14.7 to P pound.

§ 103. Formulas for Heat in Gases in regard to Pressure.

Mean specific heat, $s = S\sqrt{\frac{p}{P}}$. 1

Units of heat, $h = S\,W\sqrt{\frac{p}{P}}(T - 32°)$. 2

Temperature, $T = \frac{h}{S\,W}\sqrt{\frac{P}{p}} + 32°$. 3

Pressure of gas, $P = p\left(\frac{S\,W(T - 32)}{h}\right)^2$. 4

Example 1. What is the mean specific heat of air, heated under constant volume from a pressure $p = 14.7$ to $P = 26$ pounds to the square inch?

$$\text{Mean specific heat,} \quad s = 0.25\sqrt{\frac{14.7}{26}} = 0.188.$$

Example 2. How many units of heat are there in $W = 8$ pounds of carbonic acid, heated from 32° to $T = 450°$, and from a pressure 14.7 to $P = 20$ pounds per square inch?

$$\text{Units of heat,} \quad h = 0.221 \times 8\sqrt{\frac{14.7}{20}}(450 - 32) = 629.25.$$

Example 3. What will be the temperature of $W = 12$ pounds of air supplied with $h = 864$ units of heat, which increases the pressure from $p = 14.7$ to $P = 24$ pounds to the square inch?

$$\text{Temperature,} \quad T = \frac{864}{12 \times 0.25}\sqrt{\frac{24}{14.7}} + 32 = 323.33.$$

Example 4. What pressure will be attained by heating $W = 24$ pounds of carbonic oxide from 32° to $T = 280°$, with $h = 2400$ units of heat supplied to the gas in a closed vessel?

$$\text{Pressure of gas,} \quad P = 14.7\left(\frac{0.288 \times 24(280 - 32}{2400}\right)^2 = 8.8513.$$

In this case the pressure became less than the primitive pressure, the reason of which is that the volume was expanded in order to admit 2400 units of heat without increasing the temperatures over 280°.

§ 104. Formulas for Heat in Gases in regard to Volume.

Mean specific heat, $$s = S\sqrt{\frac{\dot{\mathfrak{V}}}{\mathfrak{V}\,x}}. \qquad 5$$

Units of heat, $$h = S\,\mathfrak{W}\sqrt{\frac{\dot{\mathfrak{V}}\,\mathfrak{V}}{x}}(T-32). \qquad 6$$

Temperature, $$\mathfrak{T} = \frac{h}{S\,\mathfrak{W}}\sqrt{\frac{x}{\dot{\mathfrak{V}}\,\mathfrak{V}}}+32. \qquad 7$$

Volume, $$\dot{\mathfrak{V}} = \frac{x}{\mathfrak{V}}\left(\frac{S\,\mathfrak{W}\,(T-32)}{h}\right)^2. \qquad 8$$

Example 5. Required the mean specific heat of hydrogen gas, heated from 32° to $\mathfrak{T}=450°$, and the volume increased 50 per cent.?

$$x = 1.8477.$$

Specific heat, $$s = 3.3\sqrt{\frac{1.5}{1\times 1.8477}} = 2.9733.$$

Example 6. How many units of heat are required to heat $\mathfrak{V}=36$ cubic feet of nitrogen gas from 32° to $\mathfrak{T}=400°$, and expand the volume to $\dot{\mathfrak{V}}=40$ cubic feet?

Units of heat, $$h = 0.275\times 0.07837\sqrt{\frac{36\times 40}{1.7463}}(400-32) = 227.75.$$

By the aid of the following table the preceding formulas and calculations can be much simplified by calling

$$y = \frac{(\mathfrak{T}-32)}{\sqrt{x}} = (\mathfrak{T}-32)\sqrt{\frac{493}{461+\mathfrak{T}}}. \qquad 9$$

The value of y is calculated for different temperatures in the table, by the aid of which the units of heat in any gas can be found by the following formulas.

$$h = y\,S\,W\sqrt{\frac{\dot{\mathfrak{V}}}{\mathfrak{V}}}. \qquad 10$$

$$h = y\,S\,\mathfrak{W}\sqrt{\dot{\mathfrak{V}}\,\mathfrak{V}}. \qquad 11$$

$$y = \frac{h}{S\,W}\sqrt{\frac{\mathfrak{V}}{\dot{\mathfrak{V}}}}. \qquad 12$$

$$y = \frac{h}{S\,\mathfrak{W}\sqrt{\dot{\mathfrak{V}}\,\mathfrak{V}}}. \qquad 13$$

Having given the weight W, volumes $\mathcal{V}$ and $\mathfrak{V}$, and the units of heat h, in any permanent gas, calculate the value of y by Formula 12 or 13, which gives the corresponding temperature of the gas in the table.

Example 11. How many units of heat are required to elevate the temperature of $\mathfrak{V}=160$ cubic feet of air from 32° to $\mathfrak{T}=480°$, and expand the volume to $\mathcal{V}=240$ cubic feet?

In the table find $y=324.29$ for 480°.

Units of heat, $h=324.29\times 0.25\times 0.08042\sqrt{160\times 240}=1277.6.$

Example 13. What will be the temperature of $\mathcal{V}=36$ cubic feet of carbonic acid heated from 32° and volume $\mathfrak{V}=24$ cubic feet, when $h=140$ units of heat has been expended on it?

$$y=\frac{140}{0.221\times 0.1233\sqrt{36\times 24}}=133.8.$$

This corresponds to a temperature $\mathfrak{T}=185°$ in the table.

DRAFT IN CHIMNEYS.

§ 105. The draft in a definite chimney depends upon the temperature of the ascending gases. The higher the temperature is, the lighter will the gases be, and consequently create a stronger draft under the fire-grate, as before explained, § 45.

The velocity of the air through the fire-grate is

$$V'=8\sqrt{H\left(1-\frac{1}{x}\right)}. \quad . \quad . \quad . \quad . \qquad 1$$

Call $$z=\left(1-\frac{1}{x}\right). \quad . \quad . \quad . \quad . \quad . \qquad 2$$

Then the velocity $$V'=8\sqrt{Hz}. \quad . \quad . \quad . \quad . \quad . \qquad 3$$

$$H=\frac{64V^2}{z}. \quad . \quad . \quad . \quad . \quad . \quad . \qquad 4$$

The value of z is calculated for different temperatures of the gases in the chimney, and is contained in column z in Table XXX.

Example 3. The height of a chimney is $H=144$ feet, and temperature of the gases $\mathfrak{T}=520°$. Required the velocity of the draft through the fire-grate? See Table XXX. for temperature 520°, which corresponds to $z=0.4977$.

$$V'=8\sqrt{144\times 4977}=67.8 \text{ feet per second.}$$

TABLE XXIX.

Horse-power of Chimneys. Formula 1, § 26, page 42.

For safety this table gives the horse-power about 25 per cent. less than may be attained in practice.

Height chimn.	Area of chimney in square feet at the top.									
	0.5	1	2	4	6	10	15	20	30	40
Feet.	HP	HP	HP	HP	HP	HP	HP	HP	HP	HP
20	3.35	6.7	13.4	26.8	40.2	67	100.5	134	201	268
25	3.7	7.4	14.8	29.6	44.4	74	111.0	148	222	296
30	4.0	8.0	16.0	32.0	48.0	80	120.0	160	240	320
35	4.25	8.5	17.0	34.0	51.0	85	127.5	170	255	340
40	4.5	9.0	18.0	36.0	54.0	90	135.0	180	270	360
45	4.75	9.5	19.0	38.0	57.0	95	142.5	190	285	380
50	5.0	10.0	20.0	40.0	60.0	100	150.0	200	300	400
55	5.2	10.4	20.8	41.6	62.4	104	156.0	208	312	416
60	5.4	10.8	21.6	43.2	64.8	108	162.0	216	324	432
65	5.6	11.2	22.4	44.8	67.2	112	168.0	224	336	448
70	5.8	11.6	23.2	46.4	69.6	116	174.0	232	348	464
75	6.0	12.0	24.0	48.0	72.0	120	180.0	240	360	480
80	6.15	12.3	24.6	49.2	73.8	123	184.5	246	369	492
85	6.35	12.7	25.4	50.8	76.2	127	190.5	254	381	508
90	6.5	13.0	26.0	52.0	78.0	130	195.0	260	390	520
95	6.65	13.3	26.6	53.2	79.8	133	199.5	266	399	532
100	6.8	13.6	27.2	54.4	82.8	136	204.0	272	414	544
110	7.1	14.2	28.4	56.8	85.2	142	213.0	284	426	568
120	7.4	14.8	29.6	59.2	88.8	148	222.0	296	444	592
130	7.65	15.3	30.6	61.2	91.8	153	229.5	306	459	612
140	7.9	15.8	31.6	63.2	94.8	158	237.0	316	474	632
150	8.15	16.3	32.6	65.2	97.8	163	244.5	326	489	652
160	8.4	16.8	33.6	67.2	100.8	168	252.0	336	504	672
170	8.65	17.3	34.6	69.2	103.8	173	259.5	346	519	692
180	8.9	17.8	35.6	71.2	106.8	178	267.0	356	534	712
190	9.2	18.2	36.4	72.8	109.2	182	273.0	364	546	728
200	9.3	18.6	37.2	74.4	111.6	186	279.0	372	558	744
210	9.5	19.0	38.0	76.0	114.0	190	285 0	380	570	760
220	9.7	19.4	38.8	77.6	116.4	194	291.0	388	582	776
230	9.9	19.8	39.6	79.2	118.8	198	297.0	396	594	792
240	10.1	20.2	40.4	80.8	121.2	202	303.0	404	606	808
250	10.3	20.6	41.2	82.4	123.6	206	309.0	412	618	824
260	10.5	21.0	42.0	84.0	126.0	210	315.0	420	630	840
270	10.65	21.3	42.6	85.2	127.8	213	319.5	426	639	852
280	10.8	21.6	43.2	86.4	129.6	216	324.0	432	648	864
290	11.0	22.0	44.0	88.0	132.0	220	330.0	440	660	880
300	11.15	22 3	44.6	89.2	133.8	223	334.5	446	669	892
310	11.35	22.7	45.4	90.8	136.2	227	340.5	454	681	908
320	11.5	23.0	46.0	92.0	138.0	230	345.0	460	690	920
330	11.65	23.3	46.6	93.2	139.8	233	349.5	466	699	932
340	11.8	23.6	47.2	94.4	141.6	236	354.0	472	708	944
350	12.0	24.0	48.0	96.0	144.0	240	360.0	480	720	960
360	12.15	24.3	48.6	97.2	145.8	243	364.5	486	729	972
370	12.3	24.6	49.2	98.4	147.6	246	369.0	492	738	984
380	12.45	24.9	49.8	99.6	149.4	249	373.5	498	747	996
390	12.6	25.2	50.4	100.8	151.2	252	378.0	504	756	1008
400	12.75	25.5	51.0	102.0	153.0	255	382.5	510	765	1020

TABLE XXX.

Physical Properties of Permanent Gases.

Temp. Fahr.	$\frac{PV}{pv}$	$\frac{T-t}{\sqrt{x}}$	$1-\frac{1}{x}$	Temp. Fahr.	$\frac{PV}{pv}$	$\frac{T-t}{\sqrt{x}}$	$1-\frac{1}{x}$	Temp. Fahr.	$\frac{PV}{pv}$	$\frac{T-t}{\sqrt{x}}$	$1-\frac{1}{x}$
$T°$	x	y	z	$T°$	x	y	z	$T°$	x	y	z
−180	0.5700	−280.9	−0.261	32	1.0000	0.0000	0.0000	82	1.1014	47.643	0.0920
−170	0.5903	−262.9	−0.306	33	1.0020	0.9990	0.0019	83	1.1034	48.552	0.0935
−160	0.6106	−245.8	−0.362	34	1.0040	1.9960	0.0039	84	1.1054	49.459	0.0954
−150	0.6308	−229.3	−0.415	35	1.0061	2.9909	0.0059	85	1.1075	50.362	0.0969
−140	0.6511	−213.2	−0.464	36	1.0081	3.9839	0.0079	86	1.1095	51.266	0.0986
−130	0.6714	−197.7	−0.511	37	1.0101	4.9750	0.0099	87	1.1115	52.168	0.0999
−120	0.6917	−187.8	−0.554	38	1.0121	5.9640	0.0118	88	1.1135	53.069	0.1019
−110	0.7120	−168.3	−0.595	39	1.0142	6.9508	0.0187	89	1.1156	53.966	0.1035
−100	0.7322	−154.3	−0.634	40	1.0162	7.9360	0.0157	90	1.1176	54.851	0.1051
−90	0.7524	−140.7	−0.671	41	1.0182	8.9192	0.0176	91	1.1196	55.760	0.1069
−80	0.7727	−127.4	−0.706	42	1.0203	9.9000	0.0195	92	1.1217	56.652	0.1083
−70	0.7930	−114.5	−0.739	43	1.0223	10.880	0.0215	93	1.1237	57.555	0.1099
−60	0.8133	−102.0	−0.770	44	1.0243	11.857	0.0234	94	1.1257	58.436	0.1118
−50	0.8336	−89.82	−0.800	45	1.0264	12.834	0.0253	95	1.1277	59.326	0.1130
−40	0.8540	−77.91	−0.829	46	1.0284	13.805	0.0272	96	1.1297	60.214	0.1149
−30	0.8742	−66.31	−0.856	47	1.0304	14.777	0.0291	97	1.1318	61.098	0.1165
−20	0.8945	−54.98	−0.882	48	1.0325	15.746	0.0315	98	1.1338	61.983	0.1179
−10	0.9148	−43.91	−0.907	49	1.0345	16.714	0.0329	99	1.1358	62.867	0.1191
0	0.9352	−33.01	−0.930	50	1.0365	17.680	0.0349	100	1.1378	63.749	0.1210
1	0.9371	−32.06	−0.933	51	1.0385	18.666	0.0365	101	1.1399	64.627	0.1227
2	0.9391	−30.96	−0.935	52	1.0406	19.606	0.0389	102	1.1419	65.506	0.1243
3	0.9411	−29.89	−0.937	53	1.0426	20.567	0.0402	103	1.1439	66.384	0.1257
4	0.9432	−28.83	−0.939	54	1.0446	21.575	0.0429	104	1.1459	67.260	0.1273
5	0.9452	−27.77	−0.942	55	1.0466	22.482	0.0444	105	1.1480	68.132	0.1288
6	0.9472	−26.72	−0.944	56	1.0487	23.463	0.0464	106	1.1500	69.005	0.1304
7	0.9492	−25.67	−0.946	57	1.0507	24.390	0.0485	107	1.1520	69.877	0.1319
8	0.9513	−24.62	−0.949	58	1.0527	25.341	0.0503	108	1.1541	70.745	0.1334
9	0.9533	−23.56	−0.951	59	1.0547	26.290	0.0521	109	1.1561	71.613	0.1349
10	0.9554	−22.51	−0.953	60	1.0567	27.260	0.0539	110	1.1581	72.481	0.1364
11	0.9577	−21.46	−0.956	61	1.0588	28.184	0.0557	111	1.1602	73.344	0.1379
12	0.9594	−20.42	−0.958	62	1.0608	29.128	0.0574	112	1.1622	74.208	0.1393
13	0.9614	−19.38	−0.960	63	1.0628	30.070	0.0592	113	1.1642	75.072	0.1408
14	0.9635	−18.34	−0.962	64	1.0649	31.010	0.0610	114	1.1663	75.929	0.1423
15	0.9655	−17.30	−0.964	65	1.0669	31.949	0.0627	115	1.1683	76.790	0.1438
16	0.9675	−16.26	−0.966	66	1.0689	32.896	0.0645	116	1.1703	77.648	0.1452
17	0.9676	−15.23	−0.966	67	1.0709	33.822	0.0662	117	1.1724	78.502	0.1469
18	0.9716	−14.22	−0.971	68	1.0720	34.770	0.0671	118	1.1744	79.358	0.1486
19	0.9734	−13.18	−0.972	69	1.0740	35.703	0.0688	119	1.1764	80.212	0.1499
20	0.9756	−12.15	−0.975	70	1.0760	36.633	0.0706	120	1.1784	81.066	0.1515
21	0.9777	−11.13	−0.977	71	1.0780	37.563	0.0723	121	1.1805	81.914	0.1528
22	0.9797	−10.11	−0.979	72	1.0811	38.470	0.0749	122	1.1825	82.764	0.1541
23	0.9817	−9.089	−0.981	73	1.0831	39.396	0.0766	123	1.1845	83.621	0.1559
24	0.9837	−8.069	−0.983	74	1.0851	40.320	0.0783	124	1.1866	84.457	0.1571
25	0.9856	−7.051	−0.985	75	1.0871	41.289	0.0800	125	1.1886	85.303	0.1586
26	0.9878	−6.031	−0.988	76	1.0892	42.160	0.0817	126	1.1906	86.148	0.1601
27	0.9898	−5.029	−0.990	77	1.0912	43.079	0.0833	127	1.1927	86.988	0.1615
28	0.9917	−4.017	−0.991	78	1.0932	43.995	0.0870	128	1.1947	87.830	0.1629
29	0.9939	−3.010	−0.994	79	1.0953	44.940	0.0869	129	1.1967	88.671	0.1642
30	0.9957	−2.005	−0.995	80	1.0973	45.822	0.0883	130	1.1987	89.510	0.1657
31	0.9979	−1.002	−0.998	81	1.0993	46.734	0.0899	131	1.2008	90.374	0.1670

TABLE XXX.

Physical Properties of Permanent Gases.

Temp. Fahr. $T°$	$\frac{PV}{pv}$ x	$\frac{T-t}{\sqrt{x}}$ y	$1-\frac{1}{x}$ z	Temp. Fahr. $T°$	$\frac{PV}{pv}$ x	$\frac{T-t}{\sqrt{x}}$ y	$1-\frac{1}{x}$ z	Temp. Fahr. $T°$	$\frac{PV}{pv}$ x	$\frac{T-t}{\sqrt{x}}$ y	$1-\frac{1}{x}$ z
132	1.2028	91.152	0.1686	182	1.3041	131.34	0.2331	232	1.4056	168.70	0.2885
133	1.2048	92.016	0.1699	183	1.3062	132.11	0.2343	233	1.4076	169.42	0.2895
134	1.2069	92.846	0.1714	184	1.3082	132.88	0.2355	234	1.4096	170.14	0.2905
135	1.2089	93.579	0.1728	185	1.3102	133.65	0.2367	235	1.4116	170.86	0.2915
136	1.2109	94.510	0.1742	186	1.3122	134.42	0.2378	236	1.4137	171.58	0.2925
137	1.2129	95.340	0.1755	187	1.3143	135.19	0.2390	237	1.4157	172.29	0.2935
138	1.2150	96.165	0.1769	188	1.3163	135.96	0.2402	238	1.4177	173.01	0.2945
139	1.2170	96.993	0.1782	189	1.3184	136.73	0.2414	239	1.4198	173.73	0.2955
140	1.2190	97.819	0.1796	190	1.3204	137.50	0.2426	240	1.4218	174.44	0.2965
141	1.2211	98.640	0.1809	191	1.3224	138.27	0.2438	241	1.4238	175.15	0.2976
142	1.2231	99.463	0.1823	192	1.3244	139.04	0.2449	242	1.4258	175.86	0.2986
143	1.2251	100.29	0.1836	193	1.3265	139.81	0.2461	243	1.4279	176.57	0.2996
144	1.2272	101.10	0.1849	194	1.3285	140.58	0.2472	244	1.4299	177.28	0.3006
145	1.2292	101.92	0.1863	195	1.3305	141.35	0.2483	245	1.4319	177.99	0.3016
146	1.2312	102.74	0.1876	196	1.3326	142.12	0.2494	246	1.4340	178.70	0.3026
147	1.2333	103.55	0.1889	197	1.3346	142.89	0.2506	247	1.4360	179.41	0.3036
148	1.2353	104.37	0.1902	198	1.3366	143.66	0.2517	248	1.4380	180.12	0.3046
149	1.2373	105.18	0.1915	199	1.3386	144.42	0.2529	249	1.4401	180.83	0.3056
150	1.2393	106.00	0.1928	200	1.3407	145.19	0.2541	250	1.4421	181.54	0.3066
151	1.2414	106.81	0.1941	201	1.3427	145.95	0.2553	251	1.4441	182.24	0.3076
152	1.2434	107.62	0.1954	202	1.3447	146.70	0.2565	252	1.4462	182.94	0.3086
153	1.2454	108.43	0.1967	203	1.3468	147.44	0.2575	253	1.4582	183.64	0.3096
154	1.2475	109.23	0.1984	204	1.3488	148.18	0.2586	254	1.4402	184.34	0.3104
155	1.2495	110.04	0.1996	205	1.3508	148.92	0.2597	255	1.4522	185.04	0.3112
156	1.2515	110.84	0.2003	206	1.3529	149.66	0.2608	256	1.4543	185.74	0.3122
157	1.2535	111.65	0.2022	207	1.3549	150.39	0.2619	257	1.4563	186.44	0.3131
158	1.2556	112.45	0.2035	208	1.3569	151.12	0.2630	258	1.4583	187.14	0.3141
159	1.2576	113.25	0.2047	209	1.3589	151.85	0.2641	259	1.4604	187.84	0.3151
160	1.2596	114.05	0.2060	210	1.3610	152.58	0.2652	260	1.4624	188.54	0.3159
161	1.2616	114.85	0.2072	211	1.3630	153.32	0.2663	261	1.4644	189.24	0.3169
162	1.2637	115.64	0.2086	212	1.3650	154.06	0.2674	262	1.4664	189.93	0.3178
163	1.2657	116.44	0.2098	213	1.3670	154.80	0.2685	263	1.4685	190.62	0.3187
164	1.2677	117.24	0.2111	214	1.3691	155.54	0.2695	264	1.4705	191.32	0.3199
165	1.2697	118.04	0.2123	215	1.3711	156.28	0.2705	265	1.4725	192.01	0.3209
166	1.2717	118.83	0.2136	216	1.3731	157.02	0.2716	266	1.4745	192.70	0.3217
167	1.2738	119.62	0.2149	217	1.3751	157.76	0.2727	267	1.4766	193.39	0.3227
168	1.2758	120.41	0.2161	218	1.3772	158.50	0.2737	268	1.4786	194.08	0.3236
169	1.2778	121.20	0.2173	219	1.3792	159.24	0.2748	269	1.4806	194.77	0.3246
170	1.2798	121.98	0.2186	220	1.3812	159.97	0.2758	270	1.4826	195.46	0.3255
171	1.2818	122.77	0.2198	221	1.3832	160.71	0.2768	271	1.4847	196.15	0.3265
172	1.2839	123.56	0.2210	222	1.3853	161.45	0.2781	272	1.4867	196.84	0.3274
173	1.2859	124.35	0.2222	223	1.3873	162.19	0.2792	273	1.4887	197.53	0.3284
174	1.2879	125.13	0.2236	224	1.3893	162.93	0.2803	274	1.4907	198.22	0.3293
175	1.2899	125.91	0.2248	225	1.3913	163.67	0.2814	275	1.4928	198.90	0.3302
176	1.2920	126.69	0.2259	226	1.3934	164.41	0.2824	276	1.4948	199.58	0.3310
177	1.2940	127.47	0.2271	227	1.3954	165.15	0.2834	277	1.4968	200.26	0.3319
178	1.2960	128.25	0.2283	228	1.3974	165.88	0.2844	278	1.4988	200.94	0.3327
179	1.2980	129.02	0.2295	229	1.3995	166.61	0.2854	279	1.5009	201.62	0.3337
180	1.3001	129.80	0.2307	230	1.4015	167.25	0.2864	280	1.5029	202.30	0.3346
181	1.3021	130.57	0.2329	231	1.4035	167.98	0.2874	281	1.5049	202.98	0.3355

TABLE XXX.

Physical Properties of Permanent Gases.

Temp. Fahr.	$\frac{PV}{pv}$	$\frac{T-t}{\sqrt{x}}$	$1-\frac{1}{x}$	Temp. Fahr.	$\frac{PV}{pv}$	$\frac{T-t}{\sqrt{x}}$	$1-\frac{1}{x}$	Temp. Fahr.	$\frac{PV}{pv}$	$\frac{T-t}{\sqrt{x}}$	$1-\frac{1}{x}$
T	x	y	z	T	x	y	z	T	x	y	z
282	1.5070	203.66	0.3363	332	1.6084	236.55	0.3781	382	1.7097	267.71	0.4150
283	1.5090	204.34	0.3372	333	1.6104	237.19	0.3788	383	1.7118	268.33	0.4157
284	1.5110	205.02	0.3381	334	1.6124	237.83	0.3796	384	1.7138	268.94	0.4164
285	1.5131	205.70	0.3390	335	1.6144	238.43	0.3804	385	1.7158	269.55	0.4171
286	1.5151	206.37	0.3399	336	1.6165	239.11	0.3811	386	1.7179	270.16	0.4179
287	1.5171	207.04	0.3407	337	1.6185	239.75	0.3819	387	1.7199	270.77	0.4185
288	1.5192	207.71	0.3416	338	1.6205	240.39	0.3827	388	1.7219	271.38	0.4192
289	1.5212	208.38	0.3425	339	1.6226	241.02	0.3836	389	1.7240	271.99	0.4199
290	1.5232	209.05	0.3433	340	1.6246	241.65	0.3845	390	1.7260	272.50	0.4206
291	1.5252	209.72	0.3442	341	1.6266	242.28	0.3852	391	1.7280	273.10	0.4212
292	1.5273	210.39	0.3458	342	1.6286	242.91	0.3859	392	1.7301	273.70	0.4219
293	1.5293	211.06	0.3459	343	1.6307	243.54	0.3868	393	1.7321	274.30	0.4226
294	1.5313	211.73	0.3468	344	1.6327	244.17	0.3875	394	1.7341	274.90	0.4232
295	1.5334	212.40	0.3476	345	1.6347	244.80	0.3882	395	1.7361	275.49	0.4239
296	1.5354	213.07	0.3485	346	1.6368	245.43	0.3889	396	1.7382	276.09	0.4246
297	1.5374	213.74	0.3493	347	1.6388	246.06	0.3897	397	1.7402	276.69	0.4252
298	1.5395	214.40	0.3501	348	1.6408	246.69	0.3906	398	1.7422	277.29	0.4259
299	1.5415	215.06	0.3510	349	1.6429	247.31	0.3913	399	1.7443	277.89	0.4265
300	1.5435	215.72	0.3518	350	1.6449	247.93	0.3920	400	1.7463	278.48	0.4272
301	1.5455	216.38	0.3527	351	1.6469	248.56	0.3928	401	1.7483	279.08	0.4279
302	1.5476	217.04	0.3539	352	1.6490	249.19	0.3935	402	1.7504	279.68	0.4285
303	1.5496	217.70	0.3548	353	1.6510	249.82	0.3942	403	1.7524	280.27	0.4292
304	1.5516	218.36	0.3556	354	1.6530	250.45	0.3950	404	1.7544	280.86	0.4298
305	1.5537	219.02	0.3584	355	1.6551	251.08	0.3957	405	1.7564	281.45	0.4305
306	1.5557	219.68	0.3573	356	1.6571	251.70	0.3964	406	1.7585	282.04	0.4314
307	1.5577	220.34	0.3581	357	1.6591	252.32	0.3971	407	1.7605	282.63	0.4320
308	1.5597	221.00	0.3589	358	1.6611	252.94	0.3979	408	1.7625	283.22	0.4325
309	1.5618	221.65	0.3597	359	1.6632	253.56	0.3986	409	1.7646	283.81	0.4332
310	1.5638	222.30	0.3605	360	1.6652	254.18	0.3993	410	1.7666	284.40	0.4340
311	1.5658	222.96	0.3614	361	1.6672	254.80	0.4001	411	1.7686	284.99	0.4346
312	1.5678	223.61	0.3622	362	1.6692	255.42	0.4008	412	1.7706	285.58	0.4353
313	1.5699	224.27	0.3630	363	1.6713	256.04	0.4015	413	1.7727	286.17	0.4359
314	1.5719	224.93	0.3638	364	1.6733	256.66	0.4022	414	1.7747	286.76	0.4365
315	1.5739	225.58	0.3646	365	1.6753	257.28	0.4029	415	1.7767	287.35	0.4372
316	1.5759	226.23	0.3654	366	1.6773	257.90	0.4036	416	1.7787	287.94	0.4378
317	1.5780	226.88	0.3662	367	1.6794	258.51	0.4043	417	1.7808	288.43	0.4384
318	1.5800	227.53	0.3670	368	1.6814	259.12	0.4051	418	1.7828	289.01	0.4391
319	1.5820	228.28	0.3678	369	1.6834	259.74	0.4058	419	1.7848	289.59	0.4397
320	1.5840	228.83	0.3686	370	1.6854	260.36	0.4065	420	1.7868	290.27	0.4403
321	1.5861	229.48	0.3694	371	1.6875	260.97	0.4072	421	1.7889	290.85	0.4410
322	1.5881	230.13	0.3702	372	1.6895	261.58	0.4079	422	1.7909	291.43	0.4416
323	1.5901	230.78	0.3710	373	1.6915	262.19	0.4085	423	1.7929	292.01	0.4422
324	1.5922	231.42	0.3718	374	1.6935	262.80	0.4096	424	1.7950	292.59	0.4428
325	1.5942	232.06	0.3726	375	1.6956	263.41	0.4103	425	1.7970	293.17	0.4434
326	1.5962	232.71	0.3734	376	1.6976	264.02	0.4110	426	1.7990	293.75	0.4441
327	1.5982	233.35	0.3741	377	1.6996	264.63	0.4116	427	1.8010	294.33	0.4447
328	1.6003	233.90	0.3749	378	1.7016	265.24	0.4123	428	1.8031	294.91	0.4453
329	1.6023	234.63	0.3757	379	1.7037	265.85	0.4130	429	1.8051	295.49	0.4459
330	1.6043	235.27	0.3765	380	1.7057	266.46	0.4137	430	1.8071	296.07	0.4465
331	1.6063	235.91	0.3773	381	1.7077	267.09	0.4144	431	1.8091	296.65	0.4471

TABLE XXX.

Physical Properties of Permanent Gases.

Temp. Fahr.	$\frac{PV}{pv}$	$\frac{T-t}{\sqrt{x}}$	$1-\frac{1}{x}$	Temp. Fahr.	$\frac{PV}{pv}$	$\frac{T-t}{\sqrt{x}}$	$1-\frac{1}{x}$	Temp. Fahr.	$\frac{PV}{pv}$	$\frac{T-t}{\sqrt{x}}$	$1-\frac{1}{x}$
$T°$	x	y	z	$T°$	x	y	z	$T°$	x	y	z
432	1.8112	297.23	0.4477	482	1.9126	325.39	0.4771	820	2.5980	488.87	0.6150
433	1.8132	297.81	0.4483	483	1.9146	325.94	0.4777	830	2.6183	493.14	0.6180
434	1.8152	298.39	0.4490	484	1.9166	326.49	0.4783	840	2.6385	497.43	0.6209
435	1.8173	298.97	0.4496	485	1.9187	327.04	0.4789	850	2.6588	501.66	0.6239
436	1.8193	299.54	0.4502	486	1.9207	327.59	0.4794	860	2.6791	505.88	0.6267
437	1.8213	300.11	0.4508	487	1.9227	328.14	0.4799	870	2.6994	510.07	0.6294
438	1.8234	300.68	0.4524	488	1.9248	328.69	0.4805	880	2.7197	514.23	0.6323
439	1.8254	301.25	0.4520	489	1.9268	329.24	0.4811	890	2.7399	518.36	0.6350
440	1.8274	301.82	0.4526	490	1.9288	329.78	0.4816	900	2.7602	522.45	0.6376
441	1.8294	302.39	0.4532	491	1.9309	330.33	0.4821	910	2.7805	526.54	0.6403
442	1.8315	302.96	0.4538	492	1.9329	330.88	0.4827	920	2.8008	530.61	0.6429
443	1.8335	303.53	0.4544	493	1.9349	331.43	0.4832	930	2.8211	534.66	0.6455
444	1.8355	304.10	0.4550	494	1.9369	331.98	0.4838	940	2.8413	538.71	0.6480
445	1.8376	304.67	0.4558	495	1.9390	332.52	0.4843	950	2.8616	542.67	0.6504
446	1.8396	305.24	0.4564	496	1.9410	333.06	0.4847	960	2.8819	546.66	0.6529
447	1.8416	305.81	0.4570	497	1.9430	333.60	0.4852	970	2.9022	550.60	0.6554
448	1.8436	306.37	0.4576	498	1.9451	334.14	0.4857	980	2.9225	554.52	0.6577
449	1.8457	306.94	0.4581	499	1.9471	334.68	0.4863	990	2.9427	558.45	0.6601
450	1.8477	307.51	0.4587	500	1.9491	335.22	0.4869	1000	2.9630	562.36	0.6624
451	1.8497	308.08	0.4593	510	1.9694	340.60	0.4921	1010	2.9833	566.24	0.6647
452	1.8518	308.65	0.4599	520	1.9898	345.95	0.4977	1020	3.0036	570.09	0.6670
453	1.8538	309.22	0.4605	530	2.0102	351.26	0.5024	1030	3.0239	573.94	0.6692
454	1.8558	309.79	0.4611	540	2.0302	356.53	0.5073	1040	3.0441	577.73	0.6714
455	1.8579	310.35	0.4617	550	2.0505	361.75	0.5120	1050	3.0644	581.53	0.6736
456	1.8599	310.91	0.4623	560	2.0708	366.93	0.5171	1060	3.0847	585.32	0.6758
457	1.8619	311.47	0.4629	570	2.0909	372.06	0.5217	1070	3.1050	589.08	0.6779
458	1.8639	312.03	0.4635	580	2.1113	377.16	0.5262	1080	3.1253	592.82	0.6799
459	1.8660	312.59	0.4641	590	2.1316	382.28	0.5308	1090	3.1455	596.54	0.6820
460	1.8680	313.15	0.4647	600	2.1519	387.20	0.5353	1100	3.1658	600.24	0.6841
461	1.8700	313.71	0.4652	610	2.1721	392.18	0.5395	1110	3.1861	603.92	0.6861
462	1.8720	314.27	0.4657	620	2.1924	397.13	0.5437	1120	3.2064	607.62	0.6880
463	1.8741	314.83	0.4663	630	2.2127	402.03	0.5481	1130	3.2267	611.27	0.6901
464	1.8761	315.39	0.4669	640	2.2329	406.89	0.5521	1140	3.2469	614.92	0.6920
465	1.8781	315.95	0.4675	650	2.2532	411.71	0.5561	1150	3.2672	618.52	0.6938
466	1.8801	316.51	0.4681	660	2.2734	416.50	0.5601	1160	3.2875	622.13	0.6957
467	1.8822	317.07	0.4686	670	2.2938	421.25	0.5640	1170	3.3078	625.73	0.6976
468	1.8842	317.63	0.4692	680	2.3141	425.98	0.5678	1180	3.3281	629.32	0.6994
469	1.8862	318.19	0.4697	690	2.3343	430.67	0.5715	1190	3.3484	632.90	0.7013
470	1.8882	318.75	0.4703	700	2.3545	435.34	0.5752	1200	3.3687	636.38	0.7031
471	1.8903	319.31	0.4709	710	2.3749	439.88	0.5789	1300	3.5714	671.08	0.7199
472	1.8923	319.87	0.4714	720	2.3952	444.52	0.5824	1400	3.7743	704.74	0.7350
473	1.8943	320.43	0.4720	730	2.4155	449.11	0.5859	1500	3.9770	737.35	0.7485
474	1.8963	320.99	0.4726	740	2.4357	453.67	0.5894	1600	4.1798	766.95	0.7608
475	1.8984	321.54	0.4731	750	2.4560	458.15	0.5928	1700	4.3826	797.49	0.7768
476	1.9004	322.09	0.4736	760	2.4763	462.60	0.5961	1800	4.5854	826.60	0.7818
477	1.9024	322.64	0.4742	770	2.4966	467.03	0.5993	1900	4.7882	854.45	0.7911
478	1.9044	323.19	0.4747	780	2.5169	471.44	0.6026	2000	4.9910	880.91	0.7996
479	1.9065	323.74	0.4752	790	2.5371	475.84	0.6058	2100	5.1938	906.76	0.8074
480	1.9085	324.29	0.4758	800	2.5574	480.24	0.6089	2200	5.3966	931.72	0.8147
481	1.9105	324.84	0.4764	810	2.5777	484.56	0.6120	2300	5.5994	957.80	0.8213

COMPRESSION AND EXPANSION OF A DEFINITE WEIGHT OF AIR.

§ 107. This subject does not yet seem to have been satisfactorily treated, either by experiments or mathematics, for which reason the following formulas and tables can be considered approximately correct only within our limit of practice. The assumption of the existence of an absolute zero at $-461°$, and that gases are still permanent at that temperature, does not appear to agree with the experiments on the compression and expansion of a definite weight of air. In order to make the exponents of the formulas of even numbers, the temperature $-343°$ is herein adopted as an ideal zero, not with assumption that this is an absolute zero, but it may be the temperature about which air condenses to liquid or freezes solid and its pressure ceases.

It is supposed in the following formulas that a definite weight of air is enclosed in a vessel, which volume can be increased or diminished without losing or gaining any weight of the air enclosed therein, and that no heat is lost or gained by conduction or radiation to or from the sides of the vessel.

$\mathfrak{V}$ = volume and t = temperature of the air to be compressed or expanded to the volume $\mathfrak{v}$ of temperature T.

Thus, when the air is compressed, the small volume is $\mathfrak{v}$ and the high temperature is T; but when the air is expanded, $\mathfrak{v}$ means the large volume and T the lowest temperature.

$$\frac{\mathfrak{v}}{\mathfrak{V}} = \left(\frac{t+343}{T+343}\right)^2. \qquad 1$$

$\mathfrak{T} = (T+343)$, the ideal temperature of the volume $\mathfrak{v}$.
$\mathfrak{t} = (t+343)$, the ideal temperature of the volume $\mathfrak{V}$.

§ 108. VOLUME AND TEMPERATURE.

$$\frac{\mathfrak{v}}{\mathfrak{V}} = \left(\frac{\mathfrak{t}}{\mathfrak{T}}\right)^2 \quad \text{and} \quad \frac{\mathfrak{V}}{\mathfrak{v}} = \left(\frac{\mathfrak{T}}{\mathfrak{t}}\right)^2. \qquad 2$$

$$\mathfrak{v} = \mathfrak{V}\left(\frac{\mathfrak{t}}{\mathfrak{T}}\right)^2. \qquad 3$$

$$\mathfrak{T} = \mathfrak{t}\sqrt{\frac{\mathfrak{V}}{\mathfrak{v}}}. \qquad 4$$

Compression of Air.

Example 3. To what volume must $\mathfrak{V}=9$ cubic feet of air of $t=62°$ be compressed in order to increase the temperature to $T=552°$?

$$\mathfrak{t}=62+343=405°. \qquad \mathfrak{T}=552+343=895°.$$

$$\text{Volume,} \quad \mathfrak{v}=9\left(\frac{405}{895}\right)^2=1.843 \text{ cubic feet.}$$

Example 4. A volume of air $\mathfrak{V}=5$ cubic inches of $t=75°$ is to be compressed to $\mathfrak{v}=0.35$ cubic inches. Required the temperature of the compressed volume?

$$\mathfrak{t}=75+343=418°.$$

$$\text{Temperature,} \quad \mathfrak{T}=418\sqrt{\frac{5}{0.35}}=1607.2.$$

$$T=1607.2-343=1264.2°, \text{ the temperature required.}$$

Expansion of Air.

Example 4. A volume of air $\mathfrak{V}=12$ cubic feet and of temperature $t=57°$ is to be expanded to $\mathfrak{v}=36$ cubic feet. Required the temperature of the expanded volume?

$$\text{Ideal temperature,} \quad \mathfrak{T}=400\sqrt{\frac{12}{36}}=230.95°.$$

$$343-231=-112°, \text{ the required temperature.}$$

Example 3. How much must air of $t=32$ be expanded in order tc reduce the temperature to $T=-80°$?

$$\mathfrak{T}=343+80=163° \quad \text{and} \quad \mathfrak{t}=343+32=375°.$$

$$\text{Volume,} \quad \mathfrak{v}=\left(\frac{375}{163}\right)^2=5.293 \text{ times the primitive volume.}$$

§ 109. PRESSURE AND TEMPERATURE.

$$\frac{P}{p}=\left(\frac{\mathfrak{T}}{\mathfrak{t}}\right)^3 \quad \text{and} \quad \frac{p}{P}=\left(\frac{\mathfrak{t}}{\mathfrak{T}}\right)^3. \quad . \quad . \quad 5$$

P = pressure at temperature $\mathfrak{T}$ or T.

p = primitive pressure at temperature $\mathfrak{t}$ or t.

The pressures mean above vacuum,

$$P = p\left(\frac{\mathfrak{T}}{\mathfrak{t}}\right)^3. \quad . \quad . \quad . \quad . \quad . \quad . \quad . \quad 6$$

$$\mathfrak{T} = \mathfrak{t}\sqrt[3]{\frac{P}{p}}. \quad . \quad . \quad . \quad . \quad . \quad . \quad . \quad 7$$

Compression of Air.

Example 6. A volume of air of $p = 14.7$ pounds pressure and of temperature $\mathfrak{t} = 52°$ is to be compressed until the temperature becomes $\mathfrak{T} = 360°$. Required the pressure of the air at that temperature?

Pressure, $$P = 14.7\left(\frac{703}{375}\right)^3 = 96.84 \text{ pounds.}$$

Example 7. A volume of air of pressure $p = 16$ pounds to the square inch and of temperature $\mathfrak{t} = 45°$ is to be compressed to $P = 80$ pounds per square inch. Required the temperature of the compressed air?

$$\mathfrak{t} = 343 + 45 = 388°.$$

Ideal temperature, $$\mathfrak{T} = 388\sqrt[3]{\frac{80}{16}} = 663.48°.$$

$$T = 663.48 - 343 = 320.48°, \text{ the temperature required.}$$

Expansion of Air.

Example 6. Air of pressure $p = 14.7$ pounds and $\mathfrak{t} = 48°$ is to be expanded until the temperature becomes $\mathfrak{T} = -12°$. Required the pressure of the expanded air?

Pressure, $$P = 14.7\left(\frac{331}{391}\right)^3 = 8.9181 \text{ pounds.}$$

Example 7. A volume of air of pressure $p = 15$ pounds and temperature $\mathfrak{t} = 80°$ is to be expanded until the pressure becomes $P = 5$ pounds to the square inch. Required the temperature of the expanded air?

$$\mathfrak{t} = 343 + 80 = 423.$$

Ideal temperature, $$\mathfrak{T} = 423\sqrt[3]{\frac{5}{15}} = 293.3°.$$

The required temperature, $\mathfrak{T} = 293.3 - 343 = 49.7°$ below Fahr. zero.

§ 110. VOLUME AND PRESSURE.

$$\sqrt{\frac{\not V}{\mathfrak{V}}} = \sqrt[3]{\frac{p}{P}} \quad \text{and} \quad \sqrt{\frac{\mathfrak{V}}{\not V}} = \sqrt[3]{\frac{P}{p}}. \quad . \quad . \quad 8$$

$$\not V = \mathfrak{V}\sqrt[3]{\left(\frac{p}{P}\right)^2}. \quad . \quad . \quad . \quad . \quad . \quad . \quad 9$$

$$P = p\sqrt[3]{\left(\frac{\mathfrak{V}}{\not V}\right)^3}. \quad . \quad . \quad . \quad . \quad . \quad . \quad 10$$

Compression of Air.

Example 9. A volume of air $\mathfrak{V} = 18$ cubic feet of pressure $p = 15$ pounds is compressed to $P = 25$ pounds to the square inch. Required the volume of the compressed air?

$$\text{Volume,} \quad \not V = 18\sqrt[3]{\left(\frac{15}{25}\right)^2} = 12.805 \text{ cubic feet.}$$

Example 10. A volume of air $\mathfrak{V} = 24$ cubic inches and $p = 15$ pounds is compressed to $\not V = 6$ cubic inches. Required the pressure of the compressed volume?

$$\text{Pressure,} \quad P = 15\sqrt{\left(\frac{24}{6}\right)^3} = 120 \text{ pounds to the square inch.}$$

Expansion of Air.

Example 9. A volume of air $\mathfrak{V} = 5$ cubic metres and of pressure $p = 1$ atmosphere is to be expanded to $P = 0.25$ of an atmosphere. Required the volume of the expanded air?

$$\text{Volume} \quad \not V = 5\sqrt[3]{\left(\frac{1}{0.25}\right)^2} = 12.6 \text{ cubic metres.}$$

Example 10. What will be the pressure of air expanded to 5 times its original volume?

$$\text{Pressure,} \quad P = \sqrt{\left(\frac{1}{5}\right)^3} = 0.299 \text{ of the original pressure.}$$

§ 111. WORK OF COMPRESSION.

The differential work of compression will be

$$\partial k = P\,\partial\mathfrak{v}, \text{ but } P = p\left(\frac{\mathfrak{V}}{\mathfrak{v}}\right)^{1.5}$$

$$\partial k = p\left(\frac{\mathfrak{V}}{\mathfrak{v}}\right)^{1.5}\partial\mathfrak{v} = p\;\mathfrak{V}^{1.5}\,\frac{\partial\mathfrak{v}}{\mathfrak{v}^{1.5}}$$

$$k = p\;\mathfrak{V}^{1.5}\int\frac{\partial\mathfrak{v}}{\mathfrak{v}^{1.5}} = p\;\mathfrak{V}^{1.5}\frac{1}{0.5\,\mathfrak{v}^{0.5}} = 2\,p\sqrt{\frac{\mathfrak{V}^3}{\mathfrak{v}}} + C.$$

When $\mathfrak{V} = \mathfrak{v}$, then $k = O$, and

$2\,p\sqrt{\frac{\mathfrak{V}^3}{\mathfrak{V}}} + C = O$, of which $C = -2\,p\;\mathfrak{V}$.

$$\text{The work } k = 2\,p\sqrt{\frac{\mathfrak{V}^3}{\mathfrak{v}}} - 2\,p\;\mathfrak{V},$$

$$\text{or } k = 2\,p\;\mathfrak{V}\left(\sqrt{\frac{\mathfrak{V}}{\mathfrak{v}}} - 1\right). \quad . \quad . \quad . \quad 11$$

Let $\mathfrak{V}$ and $\mathfrak{v}$ be expressed in cubic feet and $p = 14.7$ pounds to the square inch.

K = work in foot-pounds per cubic feet of $\mathfrak{V}$ compressed to $\mathfrak{v}$.

$$2\,p = 2 \times 144 \times 14.7 = 4233.6.$$

$$K = 4233.6\;\mathfrak{V}\left(\sqrt{\frac{\mathfrak{V}}{\mathfrak{v}}} - 1\right). \quad . \quad . \quad . \quad . \quad . \quad 12$$

Mean pressure, $P = \frac{29.4\;\mathfrak{V}}{\mathfrak{V} - \mathfrak{v}}\left(\sqrt{\frac{\mathfrak{V}}{\mathfrak{v}}} - 1\right)$, in pounds per square inch.

The work done by the atmospheric pressure in compressing the air is $144 \times 14.7\;(\mathfrak{V} - \mathfrak{v})$, which, subtracted from the gross work of compression, will remain the mechanic work.

$$k = 2116.8\left[2\;\mathfrak{V}\left(\sqrt{\frac{\mathfrak{V}}{\mathfrak{v}}} - 1\right) - \left(\mathfrak{V} - \mathfrak{v}\right)\right]. \quad . \quad . \quad 13$$

Example 12. Required the gross work of compressing $\mathfrak{V} = 16$ cubic feet of air to $\mathfrak{v} = 4$ cubic feet?

Gross work, $k = 4233.6 \times 16\left(\sqrt{\frac{16}{4}} - 1\right) = 67737.6$ foot-pounds.

Of this work $k = 2116.8\ (16 - 4) = 25401.6$ foot-pounds was done by the atmospheric pressure, leaving $k = 67737.6 - 25401.6 = 4233.6$ foot-pounds of mechanic work above that of the atmosphere.

§ 112. WORK OF EXPANDING AIR.

$\mathfrak{V}$ and $\mathfrak{v}$ are expressed in cubic feet.

K = work in foot-pounds done of expanding $\mathfrak{V}$ cubic feet of air to $\mathfrak{v}$.

$$K = 4233.6\ \mathfrak{V}\left(1 - \sqrt{\frac{\mathfrak{V}}{\mathfrak{v}}}\right). \quad . \quad . \quad . \quad 14$$

The work done against the atmospheric pressure will be

$$k' = 2116.8\ (\mathfrak{v} - \mathfrak{V}). \quad . \quad . \quad . \quad . \quad 15$$

Subtract Formula 14 from 15, and the remainder will be the work done in expanding the air—namely,

$$K' = 2116.8\left[\mathfrak{v} + \mathfrak{V}\left(2\sqrt{\frac{\mathfrak{V}}{\mathfrak{v}}} - 3\right)\right]. \quad . \quad . \quad 16$$

The following tables are calculated by the preceding formulas, as will be understood by the headings. The works K and k mean foot-pounds per cubic foot of the primitive volume $\mathfrak{V}$, expanded or compressed to $\mathfrak{v}$.

TABLE XXXI.

Compression of Air by External Force.

Volume. $\mathfrak{V}=1$.	Temp. Fahr.	Pressures. Atmosp.	lbs. per sq. in.	Works. Gross.	Mechanic.
$\mathfrak{v}$	$\mathfrak{T}$	A	P	K	k
1.00	32.	1.000	14.7	0.	0.
0.95	41.7	1.080	15.9	110.08	4.19
0.90	52.3	1.171	17.2	229.04	17.36
0.85	63.7	1.276	18.7	358.17	40.60
0.80	76.3	1.398	20.5	499.56	76.20
0.75	90.5	1.545	22.7	660.44	131.19
0.70	105.2	1.707	25.1	826.40	191.36
0.65	122.1	1.908	28.0	1077.4	336.29
0.60	141.1	2.151	31.6	1232.0	385.28
0.55	162.7	2.452	36.0	1475.0	522.39
0.50	187.3	2.828	41.5	1753.5	694.60
0.45	216.	3.313	48.7	2075.5	910.71
0.40	250.	3.953	58.1	2460.1	1190.0
0.35	291.	4.829	71.0	2922.4	1546.4
0.33	306.5	5.196	76.4	3095.2	1684.0
0.30	341.1	6.085	89.4	3495.7	2014.0
0.25	407.	8.000	117.6	4233.6	2671.0
0.20	495.5	11.18	164.3	5232.7	3539.3
0.15	624.1	17.15	252.1	6684.9	4885.6
0.125	718.	22.63	322.7	7740.7	5888.5
0.10	843.	31.63	465.	9157.3	7252.2
0.05	1334	89.44	1315.	14700	12690
0.04	1532	125.	1837.	16934	14902
0.03	1822	192.	2828.	20209	18156
0.02	2309	353.5	5196.	25703	23629
0.01	3407	1000	14700.	38102	36006

TABLE XXXII.

Expansion of Air by External Force.

Volume. $\mathfrak{V}=1$.	Temp. Fahr.	Pressures. Atmosp.	Pressures. lbs. per sq. in.	Works. Gross.	Works. Mechanic.
$\mathfrak{v}$	$\mathfrak{T}$	A	P	K'	k'
1.0	32	1.0	14.7	0.	0.
1.1	14.6	0.8668	12.74	197.03	14.65
1.2	−0.7	0.7607	11.18	368.87	54.49
1.3	−14.1	0.6747	9.918	517.94	117.1
1.4	−26.1	0.6037	8.874	655.58	119.1
1.5	−36.8	0.5443	8.	776.86	281.6
1.6	−46.5	0.4941	7.263	886.64	383.4
1.7	−54.4	0.4512	6.632	986.60	295.2
1.8	−63.5	0.4141	6.087	1078.1	615.3
1.9	−70.9	0.3818	5.612	1162.3	742.8
2.0	−77.8	0.3535	5.196	1239.6	877.2
2.25	−93.0	0.2963	4.355	1411.2	1235
2.5	−105.8	0.2530	3.719	1500.0	1676
2.75	−116.9	0.2193	3.223	1680.6	2024
3.0	−126.5	0.1924	2.828	1789.7	2444
3.25	−135.0	0.1707	2.509	1885.2	2877
3.50	−142.6	0.1527	2.244	1970.6	3322
3.75	−149.3	0.1377	2.024	2047.4	3774
4.	−155.5	0.1250	1.837	2116.8	4234
4.5	−166.2	0.1048	1.540	2237.8	5171
5.	−175.3	0.0894	1.314	2340.3	6127
6.	−189.9	0.0686	1.008	2505.4	8084
7.	−201.3	0.0540	0.793	2633.4	10067
8.	−210.4	0.0442	0.650	2736.8	12080
9.	−218.0	0.0370	0.544	2822.7	14112
10.	−224.4	0.0251	0.369	2894.8	16157

CARBONIC ACID AS A PERMANENT GAS.

§ 113. When carbonic acid is not in contact with its liquid, the relation between volume and pressure behaves like that of a permanent gas, and its ideal zero is about -200 centigrade.

The latest and most reliable experiments on carbonic acid as a permanent gas have been made by Dr. Andrews, from which experiments the following formulas are deduced both in centigrade and Fahrenheit's scales of temperature.

$\mathcal{V}$ = volume of carbonic acid gas of temperature $\mathcal{T}$ centigrade, and of pressure A in atmospheres, compared with the volume at zero centigrade and under atmospheric pressure.

t = Fahrenheit temperature, and

P = pressure in pounds per square inch above vacuum.

Formulas for Centigrade Scale.

Volume, $$\mathcal{V} = \frac{1}{A} + \frac{\mathcal{T} - 1.4A}{200A}. \quad 1$$

Temperature, $$\mathcal{T} = A\,(200\mathcal{V} - 1.4) - 200. \quad 2$$

Pressure, $$A = \frac{\mathcal{T} + 200}{200\mathcal{V} - 1.4}. \quad 3$$

Formulas for Fahrenheit Scale.

Volume, $$\mathcal{V} = \frac{300 + t}{22.45P}. \quad 4$$

Temperature, $$t = 22.45P\mathcal{V} - 300. \quad 5$$

Pressure, $$P = \frac{300 + t}{22.45\mathcal{V}}. \quad 6$$

The volume corresponding to $\mathcal{T} = O$ and $A = 1$, formula 1, should be the unit 1 instead of 0.993 as shown in the table; but the course of Dr. Andrew's experiments indicate that the primitive volume had probably been 0.993. The error is only 0.007, which is corrected in formula 4.

TABLE XXXIII.

Volume of Carbonic Acid Gas of Different Temperatures and Pressures.

Temperatures.		Pressure A in Atmospheres.					
Fahr.	Cent.	1	10	20	30	40	50
t	T	V	V	V	V	V	V
32	0	0.993	0.093	0.0430	0.02633	0.01800	0.013
50	10	1.044	0.098	0.0455	0.02800	0.01925	0.014
68	20	1.098	0.103	0.0480	0.02966	0.02050	0.015
86	30	1.148	0.108	0.0505	0.03133	0.02175	0.016
104	40	1.198	0.113	0.0530	0.03300	0.02300	0.017
120	50	1.248	0.118	0.0555	0.03466	0.02425	0.018
140	60	1.298	0.123	0.0580	0.03633	0.02550	0.019
158	70	1.348	0.128	0.0605	0.03800	0.02675	0.020
176	80	1.398	0.133	0.0630	0.03966	0.02800	0.021
194	90	1.448	0.138	0.0655	0.04133	0.02925	0.022
212	100	1.498	0.143	0.0680	0.03400	0.03050	0.023
230	110	1.548	0.148	0.0705	0.04466	0.03175	0.024
248	120	1.598	0.153	0.0730	0.04633	0.03300	0.025
266	130	1.648	0.158	0.0755	0.04800	0.03425	0.026
284	140	1.698	0.163	0.0780	0.04966	0.03550	0.027
302	150	1.748	0.168	0.0805	0.08050	0.03675	0.028

CARBONIC ACID AS A VAPOR.

§ 114. When carbonic acid evaporates from or condenses to liquid, the relation between temperature and pressure behaves like that of a vapor, and its ideal zero is at about $-260°$ Fahr.

The yet most reliable experiments on carbonic acid vapor have been made by Pelouze and Faraday, from which experiments the following formulas and table are deduced—namely,

T = temperature Fahrenheit of the liquid or vapor of carbonic acid.

A = pressure in atmosphere.

P = pressure in pounds per square inch above vacuum.

Pressure atmos., $$A = \frac{(T+260)^4}{208513600}. \qquad 7$$

Logarithm, 8.3191344.

Pressure lbs., $$P = \frac{(T+260)^4}{1421700}. \qquad 8$$

Logarithm, 7.1527888.

Temperature, $T = 120.17\sqrt[4]{A} - 260.$ . . . 9

Temperature, $T = 61.404\sqrt[4]{P} - 260.$ . . . 10

It appears from the above formulas that liquid carbonic acid freezes to solid at the low temperature − 260°. The freezing point of liquid carbonic acid is variously given by different authors, of which Olmstead says − 85°, but Faraday experimented with liquid carbonic acid at − 148° without it freezing.

TABLE XXXIV.

Carbonic Acid Vapor, Pressure and Temperature.

Fahr. Temp. T.	Pressures. Atm. A.	lbs. P.	Fahr. Temp. T.	Pressures. Atm. A.	lbs. P.	Fahr. Temp. T.	Pressures. Amt. A.	lbs. P.
−260	0	0	−85	4.5	66.15	88	70	1029
−192	0.1	1.47	−81	5	73.5	99	80	1176
−180	0.2	2.94	−72	6	88.2	110	90	1323
−171	0.3	4.41	−65	7	102.9	120	100	1470
−164	0.4	5.88	−58	8	117.6	129	110	1617
−159	0.5	7.35	−52	9	132.3	138	120	1764
−154	0.6	8.82	−47	10	147	146	130	1911
−150	0.7	10.29	−36	12	176.4	153	140	2058
−146	0.8	11.76	−24	15	220.5	160	150	2205
−143	0.9	13.23	− 6	20	294	167	160	2352
−140	1	14.7	+ 9	25	267.5	174	170	2499
−127	1.5	22.05	21	30	441	180	180	2646
−117	2	29.4	32	35	514.5	186	190	2739
−109	2.5	36.75	42	40	588	192	200	2940
−102	3	41.1	51	45	661.5	197	210	3087
− 96	3.5	51.45	59	50	735	207	220	3234
− 90	4	58.8	74	60	882	212	238	3500

STEAM OR AQUEOUS VAPOR.

§ 115. Water under atmospheric pressure evaporates at ordinary temperatures under the boiling point; but that evaporation takes place only on the surface in contact with the air.

When the temperature of the water is elevated to or above that of the boiling point, then evaporation takes place in any part of the water where the temperature is so elevated.

The temperature of the boiling point depends upon the pressure on the surface of the water.

P = pressure in pounds per square inch above vacuum on the surface of the water.

T = temperature Fahr. of the boiling point.

$$T = 200\sqrt[6]{P} - 101. \quad . \quad . \quad . \quad . \quad . \quad 1$$

$$P = \left(\frac{T+101}{200}\right)^6. \quad . \quad . \quad . \quad . \quad . \quad 2$$

Example 1. At what temperature will water boil under a pressure of $P = 8$ pounds to the square inch?

This is under a vacuum of $14.7 - 8 = 6.7$ pounds to the square inch.

Temperature, $\quad T = 200\sqrt[6]{8} - 101 = 181.8°.$

Example 2. What pressure is required to elevate the temperature of the boiling point of water to $T = 330°$?

Pressure, $\quad P = \left(\frac{330° + 101}{200}\right)^6 = 100$ pounds.

The temperature of the boiling point is the same as that of the steam evaporated under the same pressure.

Supposing the above formulas to be correct, the ideal zero of aqueous vapor should be at $-101°$ Fahr., or at the temperature 101° below Fahr. zero, there is no pressure of the vapor; that is, the force of attraction between the atoms is equal to the force of expansion by heat.

LATENT HEAT OF STEAM.

§ 116. One pound of water heated under atmospheric pressure, from 32° to 212°, requires 180.9 units of heat. If more heat is supplied, steam will be generated without elevating the temperature until all the water is evaporated, which requires 1146.6 units of heat, and the steam volume will be 1740 times that occupied by the water at 32°. Then, 1146.6 − 180.9 = 965.7 units of heat in the steam which have not increased its temperature. This is what is called *latent heat,* because it does not show as temperature, but is the heat consumed in performing the work of steam.

One cubic foot of water at 32° weighs 62.387 pounds, if heated to the boiling point 212°, requires 62.387 × 180.9° = 11285.8 units of heat, and if evaporated to steam under atmospheric pressure, requires 62.387 × 1146.6 = 71532.9 units of heat, of which 71532.9 − 11285.8 = 60247.1 will be latent. It is this latent heat which generated 1740 cubic feet of steam from the cubic feet of water.

The work accomplished by that latent units of heat against the atmospheric pressure will be

$$K = 144 \times 14.7 \times (1740 - 1) = 3681115 \text{ foot-pounds.}$$

Foot-pounds per unit of heat, $$J = \frac{3681115}{60247.1} = 61.1.$$

The heat expended in elevating the temperature of the water from 32° to 212° is not realized as work.

VOLUME OF WATER.

§ 117. Water, like other liquids, expands in heating and contracts in cooling, with the exception that in heating it from 32° to 40° it contracts, and expands in heating from 40° upwards. The greatest density or smallest volume of water is therefore at 40° Fahr.

The most reliable experiments made on this subject are probably those of Kopp, by which the greatest density of water is indicated to be between 39° and 40°, or nearer 39°; but however accurate these experiments might have been made, it is impossible without the aid of mathematics to determine correctly the temperature of the greatest density because the curve tangents the abcissa at that point.

The writer has treated Kopp's experiments with very careful mathematical and graphical analysis, the result of which located the greatest density of water at 40°.

The formula for volume of water deduced from Kopp's experiments is

$$v = 1 + \frac{(t - 40)^2}{1400\,t + 398500} \quad . \quad . \quad . \quad 1$$

The volume deduced from the same experiments, but with the assertion that the greatest density of water is at 39°, will be

$$\mathfrak{V} = 1 + \frac{(t - 39)^2}{1400\, T + 405400} \quad . \quad . \quad . \quad . \quad 2$$

The Formula 1 is the most correct.

LATENT AND TOTAL HEAT IN WATER FROM 32°.

§ 118. When water expands it absorbs heat, which is not indicated as temperature, but remains latent.

l = latent heat per pound of water heated from 32°.

$\mathfrak{V}$ = volume per Formula 1.

t = temperature of the water.

h = total units of heat per pound of water heated from 32°.

Latent heat, $l = 0.1\, t\, (\mathfrak{V} - 1)$. 3

Total heat, $h = 0.1\, t\, (\mathfrak{V} + 9) - 32$. 4

Pounds per Cubic foot.		Cubic Feet per Pound.	
$\mathfrak{P} = \frac{62.388}{\mathfrak{V}}$ . . .	5	$\mathfrak{C} = \frac{\mathfrak{V}}{62.388}$ . . .	7
$\mathfrak{P} = \frac{1}{\mathfrak{C}}$	6	$\mathfrak{C} = \frac{1}{\mathfrak{P}}$	8

The latent heat in water heated from 32° to 40° is negative; that is, the water indicates more temperature than units of heat imparted to it. The volume at 32° is 1.000156, and the heat required to raise the temperature of one pound of water from 32° to 40° or 88° are $0.999844 \times 8 = 7.99875$ units.

The heat required to raise the temperature of one pound of water from 32° to 212° or 180° are 181 units. The heat required to raise water from 32° to 350° or 318° are 322 units, or 4 units more than the increase of temperature.

LATENT AND TOTAL UNITS OF HEAT IN STEAM.

§ 119. The unit of heat required to elevate the temperature of one pound of water of 32° to the boiling point and evaporate it to saturated steam of temperature T is

Units of heat, $H = 1082 + 0.305\, T$. 1

Latent heat, $L = 1082 + 0.305\, T - [0.1\, T\, (\mathfrak{V} + 9) - 32]$.

$L = 1114\, T\, (0.595 - 0.1\, \mathfrak{V})$. 2

The Formula 1 is given by Regnault. The author has reason to believe that the formula for units of heat in steam evaporated from water heated from 32° should be

$$\text{per cubic foot } H' = 2.8\ P = 2.8\left(\frac{\mathfrak{T}+101}{200}\right)^6 \quad . \quad . \quad . \quad 3$$

$$\text{per pound,} \quad H = \frac{2.8\ P}{\mathfrak{V}} = \frac{2.8}{\mathfrak{V}}\left(\frac{\mathfrak{T}-101}{200}\right)^6 \quad . \quad . \quad 4$$

The latent heat in steam by the new Formulas 3 and 4 should be

$$\text{Per cubic foot,} \quad L' = 2.8\ P - \mathfrak{V}\ \mathfrak{T}. \quad . \quad . \quad . \quad . \quad 5$$

$$\text{Per pound,} \quad L = \frac{2.8\ P}{\mathfrak{V}} - \mathfrak{T} \quad . \quad . \quad . \quad . \quad . \quad . \quad 6$$

This includes also the latent heat in the water at the boiling point, which is $l = 0.1\ t\ (\mathfrak{V}-1)$.

The thermo-dynamic equivalent per unit of latent heat will be

$$J = \frac{144\ P\ (\mathfrak{V}-1}{2.8\ P - \mathfrak{V}\ \mathfrak{T}} \quad . \quad . \quad . \quad . \quad 7$$

§ 120. The combination of the Regnault formula for units of heat with the Fairbairn formula for volume of steam does not give a constant thermo-dynamic equivalent of heat, which it ought to do, and therefore either or both the formulas are defective. The arithmetical ratio 0.305 $\mathfrak{T}$ in Regnault's formula cannot be correct, for the reason that the pressure increases as the sixth power of the temperature, and the volume decreases nearly as the cube of the temperature.

The thermo-dynamic equivalent of heat in saturated steam according to Formula 3 will be

$$J = \frac{144\ P}{2.8\ P} = 51.5, \text{ a constant number.}$$

That is to say, one or each unit of heat in saturated steam of any pressure, but without expansion, generates 51.5 foot-pounds of work. This equivalent, multiplied by 1 + hyperbolic logarithm for expansion, gives the thermo-dynamic equivalent, which can be realized by steam-power.

It has been explained (§ 10) that the steam-pressure is inversely as the expansion, which rule is sufficiently correct within our limit of practice; but when the temperature of aqueous vapor is reduced to the ideal zero—101 Fahr.—its pressure will be 0; that is, the expansive force of the heat is equal to the force of attraction between the atoms of the vapor. The vapor at that temperature will maintain a constant volume without being enclosed in a vessel.

The total heat per pound of steam, Formula 4, is nearly constant for all pressures and temperatures, differing only by the latent heat in the water heated from 32° to the boiling point under the pressure *P*.

DRYNESS OR HUMIDITY OF STEAM.

§ 121. We have yet no reliable means by which to determine correctly the dryness or humidity of steam, the knowledge of which is of great importance in steam engineering.

A steam-engine supplied with over-saturated steam does not transmit the full power due to the consumption of fuel, and thus the rate of evaporation is not a correct measure of the power or steaming capacity of the boiler.

The best means yet at our disposal by which to measure the quality of the steam working an engine is to compare the steam-volume passed through the cylinder with that due to the water evaporated in the same time, but we have yet no reliable data as to the volume of steam compared with that of its water. The experiments of Fairbairn and Tate were made on a very small scale and by apparatus which did not admit of delicate measurements, and operating so widely different from that of a steam-boiler that we have reason to doubt the correctness of the steam-volume deduced therefrom; nor does that volume for different pressures agree with the law of expansion of steam—namely, that the volume is inversely as the pressure.

We know the specific gravity of steam at 212°, which, compared with that of water at 32°, makes the steam-volume at 212° = 1730 times that of water at 32°. We also know that one volume of water at 32° resolved into its elements, oxygen and hydrogen, gases heated under atmospheric pressure to 212°, makes 2610.66 volumes of gas, of which there are 870.22 volumes of oxygen and 1740.44 volumes of hydrogen.

§ 122. When the elements are again chemically combined from gas to vapor, the volume of hydrogen takes up the volume of oxygen, leaving only 1740.44 volumes of vapor, which is probably the correct volume of steam at 212°. If the volume of steam increases as the pressure increases, the steam volume at any pressure would simply be $\mathfrak{V} = 25584.468 : P$; but the decrease of volume is accompanied with an increase of temperature which expands the volume in the same ratio as the volume of water is increased for the same difference of temperature.

Call the volume of water $=1$ at 40°, then for any other temperature, according to Copp's experiments, the volume will be

$$\mathfrak{V}=1+\frac{(\mathfrak{t}-40)^2}{1400\,\mathfrak{t}+398500}. \quad . \quad . \quad . \quad . \quad 1$$

At 212° the volume of water is 1.0426. Therefore the steam volume at any pressure and temperature should be

$$\mathcal{V}=\frac{25584.468}{1.0426\,P}\left(1+\frac{(\mathfrak{T}-40)^2}{1400\,\mathfrak{T}+398500}\right). \quad . \quad . \quad . \quad 2$$

The temperature of water and steam being alike, the

$$\text{Steam volume,} \qquad \mathcal{V}=\frac{24539\,\mathfrak{V}}{P}. \quad . \quad . \quad . \quad . \quad 3$$

TABLE XXXV.

Comparison of Volume and Temperature of Steam at Different Pressures.

Steam-pressure.	Volume of Steam.		Temperature of Steam.	
	Fairbairn.	Nystrom.	Regnault.	Nystrom.
14.7	1641.5	1740	212°	212°
25	984.23	1035	240.07	241.0
50	508.29	527.2	280.89	282.8
75	348.15	355.8	307.42	309.8
100	267.80	269.4	327.6	329.9
150	187.26	181.8	358.4	360.0
200	146.93	138	381.8	382.6
300	106.54	94.22	417.7	416.5
400	86.33	71.19	445.1	441.9

Comparison of Fairbairn's experiments and formulas with the author's steam volume:

	Pres. $P=60.6$	Pres. $P=8$	Pres. $P=4.7$
By Fairbairn's formula.....	$\mathcal{V}=428$	$\mathcal{V}=2985$	$\mathcal{V}=4900$
By Fairbairn's experiment	$\mathcal{V}=432$	$\mathcal{V}=3046$	$\mathcal{V}=4914$
By the author's formula....	$\mathcal{V}=437$	$\mathcal{V}=3150$	$\mathcal{V}=5336$

The Regnault experiments on temperature and pressure of steam gave widely different results, of which an average was adopted, and it was attempted to set up a formula to follow the average curve, which was found impossible, for which reason different formulas were set up for different parts of the irregular curve.

The formula herein adopted gives a regular curve which sweeps the whole range of the Regnault experiments, and it coincides in several places with the irregular or average curve.

The volume of one pound of steam in cubic feet will be

$$\mathfrak{E} = \frac{393.333\ \mathfrak{V}}{P}.$$

The steam volume formula by Fairbairn and Tate is

$$\mathfrak{V} = 25.62 + \frac{49513}{I + 0.72}. \quad . \quad . \quad . \quad . \quad 4$$

I = inches of mercury. That is to say, the steam volume cannot be reduced below 25.62.

For very high pressures we can omit the fraction 0.72 and insert 2.0372 P for I—namely,

$$\mathfrak{V} = 25.62 + \frac{49513}{2.0372\,P} = 25.62 + \frac{24304}{P}. \quad . \quad . \quad . \quad 5$$

When the steam-pressure is $P = 24304$ pounds to the square inch, the volume should be 26.62.

The temperature corresponding to this pressure is

$$T = 200\sqrt[6]{24304} - 101 = 975° \text{ Fahr.} \quad . \quad . \quad . \quad 6$$

The volume of water at this temperature will be

$$\mathfrak{V} = 1 + \frac{(975 - 40)^2}{1400 \times 975 + 398500} = 1.64. \quad . \quad . \quad . \quad 7$$

Then $26.62 - 1.64 = 25$ volumes, of steam pressure $P = 24304$, which cannot be materallly reduced by additional pressure, because an increase of pressure would only affect the decimals of that volume. The reason why the water volume is subtracted from that of the steam, is that the water volume is considered to be the limit to which that of steam can be reduced.

It will be noticed that Fairbairn's experimental numbers, 24304 and 25.62, agree nearly with the writer's numbers, 24539 and 25, which fact deserves consideration.

Messrs. Fairbairn and Tate omitted the consideration of expansion of water, for which reason they were obliged to add the empirical constant 25.62 in their formula.

The above argument proves conclusively that the steam volume experiments, as well as the formula of Fairbairn and Tate, cannot be relied upon, and they do not agree with the law of expansion of vapors.

The object of this paragraph is to determine the dryness or humidity of steam, for which purpose the volume due to the evaporation should be compared with the volume of steam passing through the steam cylinder.

W = cubic feet of water at 32°, evaporated during N revolutions of the engine.

$\mathfrak{V}$ = Steam volume compared with that of its water at 32°.

Q = cubic feet of steam passing through the engine or cylinder at each revolution or double stroke of the piston.

N = total number of revolutions of the engine in the time W cubic feet of water is evaporated.

$\%$ = per centage of water in the steam.

$\mathfrak{v}$ = volume of water at the temperature of the steam.

$$\% = \frac{100}{\mathfrak{v}}\left(1 - \frac{Q\,N}{W\,\mathfrak{V}}\right)$$

The steam-piston and valves must be perfectly tight, and the capacity of the steam-ports and clearance of piston must be included in Q.

§ 123. In the ordinary engine the admittance of steam is generally cut off before the piston has reached the end of the stroke, in which case the steam volume Q must be determined from the indicator diagram, as follows:

Measure the steam-pressure p on the diagram where the expansion curve begins to be regular. The steam volume $\mathfrak{V}$ corresponding to this pressure must be used in the formula. Measure the distance Q in feet, which, multiplied by the area of the piston in square feet, is the cubic capacity of the steam, to which add the capacity of the clear-

Fig. 5.

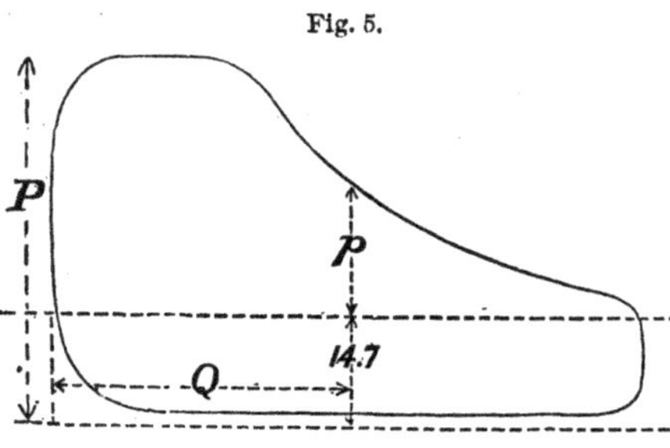

ance and steamport, and the sum is Q. This measurement must be made for both sides of the piston.

The steam-pressure should be kept as constant as possible during the experiment; but in a long run it is difficult, if not impossible, to keep it stationary, for which a mean-pressure must be determined, as follows:

The expansion being constant during the operation and the steam-pressure by gauge, noted at short and regular intervals of time, and the mean-pressure represented by p''.

p' = steam-pressure by gauge at the time the pressure p is taken on the diagram.

p''' = mean-pressure for the volume $\mathcal{V}$ in the formula.

$$p''' : p'' = p' : p \qquad \text{and} \qquad p''' = \frac{p' p''}{p}.$$

Small steam-engines ought to be constructed for the purpose of measuring the volume, dryness or humidity of steam. The slide valve in such an engine should have no lap or lead on the steam and exhaust ports, so that the full capacity of the cylinder, including clearance and steamport, would be the correct measure of the steam volume for each stroke. The cylinder and short steam-pipe could be well covered with felt, so that the pressure in the boiler would correspond to the volume $\mathcal{V}$ in the engine.

The exhaust steam could be condensed in a surface condenser and the water measured independent of the evaporation in the boiler. Such an engine could be temporarily attached to any boiler for the purpose of testing its quality of steam, and the properties of superheated steam, which are yet not well understood.

SUPERHEATING STEAM.

§ 124. When steam is superheated after generated in the water, the relation between temperature and pressure will remain the same as if the same steam had been evaporated at the same temperature as that to which it is superheated as long as it is in contact with the water. When steam is shut off from the water from which it is generated and then superheated, the relation between temperature and pressure will still remain the same as for saturated steam, provided the volume is not increased to or over 50 per cent.

When steam is superheated above the temperature and pressure due to saturated steam, and the volume is increased, the hydrogen is not capable of holding all the oxygen in its own volume; but part of the vapor is converted into gas until the volume is increased 50 per cent., when all the vapor is converted into gas. For instance, if four cubic feet of steam is superheated under constant pressure until its volume becomes six or more cubic feet, that volume will then not be vapor but a gas which may be exploded by ignition (?) In the ordinary use of steam it is never so superheated, but is always in contact with water which prevents its conversion into gas, and it requires a temperature above ignition about 600° to ignite it to explosion.

When the steam is superheated to gas it obeys the formulas for permanent gases already explained.

When steam is passed through and allowed to expand in iron tubes heated to a dull red heat, say 800°, the steam is resolved into its elements, the oxygen being taken up by the hot iron and the hydrogen gas passing off without explosion.

A definite volume of saturated steam, superheated in a closed vessel without water, will obey the formula

$$T = 200\sqrt[6]{P} - 101,$$

until the primitive pressure is increased 50 per cent., when the steam becomes a gas and obeys the formulas for permanent gases above that pressure and temperature; but being enclosed in a vessel the volume remains constant.

For instance, a volume of steam of pressure $P = 40$ pounds to the square inch, which corresponds to a temperature of

$$T = 200\sqrt[6]{40} - 101 = 268.87°,$$

is superheated under constant volume until the pressure becomes $P = 60$, the temperature will be

$$T = 200\sqrt[6]{60} - 101 = 294.7° ;$$

the steam is then a gas of $\frac{2}{3}$ volumes of hydrogen and $\frac{1}{3}$ of oxygen.

W = weight in pounds of the saturated steam superheated.

The specific heat of steam gas at 32° under atmospheric pressure is $3.3 + 0.23 = 3.53$.

The units of heat h required to superheat W pounds of saturated steam of pressure p and temperature t to pressure P and temperature T will be

$$h = 3.53\ W\sqrt{\frac{p}{P}}\,(T - t).$$

P and p both mean absolute pressures above vacuum, and the superheating accomplished without the steam being in contact with water.

$\mathfrak{V}$ = volume of the saturated steam of pressure p.

$\mathfrak{v}$ = volume of the superheated steam of pressure P.

The saturated steam becomes a perfect gas when superheated so that

$$\frac{p}{P}=\frac{\mathfrak{v}}{1.5\,\mathfrak{V}}, \text{ or when } \frac{1.5\,p}{P}=\frac{\mathfrak{v}}{\mathfrak{V}}.$$

Example. How many units of heat are required to superheat $W=3$ pounds of saturated steam of pressure $p=40$ and temperature $t=268.87°$ to a perfect gas of pressure $P=60$ and temperature $T=294.7°$?

$$\text{Units of heat } h=3.53\times 3\sqrt{\frac{40}{60}}\,(294.7-268.87)=223.34.$$

The same weight of steam raised from $p=40$ to $P=60$ of saturated steam would require only 28 units of heat, but the steam-volume which is constant in the preceding example would in this latter case be one-third less. Then $223-28=195$ units of heat expended in converting the vapor into gas and in expanding the volume 50 per cent.

It would therefore appear that there is no gain, but rather a loss, in superheating steam without contact with water for motive-power.

The expansive property of vapor generates much more power than does that of steam-gas. But when steam is to a limited extent superheated in contact with water, the expansive property is not impaired, and the water which may be carried along with the steam, is evaporated by the superheating; and thus there is a considerable gain by superheating steam, particularly when the superheating is done by the gases of combustion after having passed the water-heating surfaces. Steam-gas is very injurious to the sides and packing-rings in the cylinder; it creates more friction and is more difficult to condense than steam-vapor.

NEW TABLES FOR WATER AND STEAM.

§ 125. The following tables of properties of water and steam have been calculated by the preceding new formulas, which are considered more correct than the old ones. The meaning of each column is explained in its heading.

In the first two water-tables the pressure of the vapor in pounds per square inch is contained in the last column, of which $+P$ denotes the absolute pressure above vacuum, and $-p$ the pressure under that of the atmosphere, which is the vacuum.

TABLE XXXVI.—Properties of Water.

Temperature. Centig.	Temperature. Fahr.	Volume. Wat. = 1 at 40°.	Weight per cubic foot.	Bulk. cubic feet per lb.	Units of heat. per lb.	Units of heat. pr. c. ft.	Pressure of vapor. Absol.	Pressure of vapor. under at.
t	T	V	W	Є	h.	h'.	+ P.	— p.
0.	32	1.000109	62.3871	0.0160304	0.00000	0.0000	0.0864	–14.614
0.55	33	1.000077	62.3830	0.0160299	1.00000	62.383	0.0904	–14.610
1.11	34	1.000055	62.3842	0.0160295	2.00000	124.77	0.0945	–14.606
1.66	35	1.000035	62.3859	0.0160292	3.00001	187.16	0.0988	–14.601
2.22	36	1.000020	62.3868	0.0160290	4.00003	249.55	0.1033	–14.597
2.77	37	1.000009	62.3875	0.0160288	5.00006	311.99	0.1079	–14.592
3.33	38	1.000003	62.3876	0.0160288	6.00010	374.33	0.1127	–14.587
3.88	39	1.000001	62.3879	0.0160287	7.00015	436.72	0.1176	–14.582
4.44	40	1.000000	62.3880	0.0160287	8.00022	499.12	0.1228	–14.577
5.00	41	1.000003	62.3878	0.0160288	9.00030	561.51	0.1281	–14.571
5.55	42	1.000016	62.3873	0.0160290	10.00040	623.89	0.1336	–14.566
6.11	43	1.000034	62.3859	0.0160292	11.00051	686.28	0.1393	–14.561
6.66	44	1.000053	62.3847	0.0160295	12.00065	748.66	0.1452	–14.555
7.22	45	1.000077	62.3832	0.0160299	13.00081	811.03	0.1513	–14.549
7.77	46	1.000101	62.3815	0.0160304	14.00098	879.40	0.1576	–14.542
8.33	47	1.000136	62.3797	0.0160308	15.00132	935.70	0.1642	–14.536
8.88	48	1.000171	62.3774	0.0160314	16.00140	997.77	0.1709	–14.529
9.44	49	1.000211	62,3749	0.0160321	17.00165	1060.0	0.1780	–14.522
10.00	50	1.000254	62.3722	0.0160328	18.00192	1122.8	0.1852	–14.515
10.55	51	1.000302	62.3692	0.0160335	19.00222	1185.1	0.1927	–14.507
11.11	52	1.000353	62.3660	0.0160344	20.00255	1248.0	0.2004	–14.499
11.66	53	1.000408	62.3626	0.0160352	21.00292	1310.1	0.2084	–14.491
12.22	54	1.000468	62.3589	0.0160362	22.00329	1372.3	0.2166	–14.483
12.77	55	1.000531	62.3549	0.0160372	23.00370	1434.3	0.2252	–14.475
13.33	56	1.000597	62.3508	0.0160383	24.00415	1496.4	0.2339	–14.466
13.88	57	1.000668	62.3464	0.0160394	25.00462	1558.6	0.2430	–14.457
14.44	58	1.000740	62.3419	0.0160405	26.00513	1620.9	0.2524	–14.448
15.00	59	1.000819	62.3370	0.0160418	27.00568	1683.2	0.2621	–14.438
15.55	60	1.000901	62.3319	0.0160431	28.00626	1745.5	0.2720	–14.428
16.11	61	1.000986	62.3266	0.0160445	29.00687	1807.8	0.2824	–14.418
16.66	62	1.001075	62.3211	0.0160459	30.00752	1870.1	0.2930	–14.407
17.22	63	1.001167	62.3153	0.0160474	31.00821	1932.4	0.3040	–14.396
17.77	64	1.001262	62.3094	0.0160489	32.00894	1994.4	0.3153	–14.385
18.33	65	1.001362	62.3032	0.0160505	33.00970	2056.6	0.3269	–14.373
18.88	66	1.001464	62.2968	0.0160522	34.01051	2118.7	0.3389	–14.361
19.44	67	1.001570	62.2902	0.0160539	35.01136	2180.8	0.3513	–14.349
20.00	68	1.001680	62.2834	0.0160556	36.01224	2242.9	0.3640	–14.336
20.55	69	1.001793	62.2763	0.0160575	37.01377	2305.0	0.3771	–14.323
21.11	70	1.001909	62.2692	0.0160592	38.01415	2367.1	0.3906	–14.309
21.66	71	1.002028	62.2618	0.0160612	39.01516	2429.2	0.4045	–14.296
22.22	72	1.002151	62.2541	0.0160632	40.01622	2491.2	0.4188	–14.281
22.77	73	1.002277	62.2463	0.0160652	41.01733	2553.2	0.4336	–14.266
23.33	74	1.002406	62.2383	0.0160673	42.01848	2615.2	0.4487	–14.251
23.88	75	1.002539	62.2300	0.0160694	43.01968	2677.1	0.4644	–14.236
24.44	76	1.002675	.62.2216	.0.0160716	44.02092	2739.2	0.4804	–14.220
25.00	77	1.002814	62.2130	0.0160738	45.02222	2801.0	0.4970	–14.203
25.55	78	1.002956	62.2042	0.0160761	46.02356	2862.8	0.5139	–14.186
26.11	79	1.003101	62.1952	0.0160784	47.02495	2924.6	0.5314	–14.169
26.66	80	1.003249	62.1860	0.0160808	48.02640	2985.4	0.5493	–14.151
27.22	81	1.003400	62.1766	0.0160832	49.02789	3048.2	0.5677	–14.132
27.77	82	1.003554	62.1671	0.0160857	50.02944	3111.0	0.5868	–14.113
28.33	83	1.003711	62.1574	0.0160882	51.03104	3172.8	0.6063	–14.093
28.88	84	1.003872	62.1474	0.0160908	52.03269	3234.4	0.6264	–14.074
29.44	85	1.004035	62.1373	0.0160934	53.03439	3296.2	0.6470	–14.053
30.00	86	1.004199	62.1272	0.0160960	54.03615	3358.2	0.6681	–14.032
30.55	87	1.004370	62.1166	0.0160987	55.03797	3418.7	0.6898	–14.010
31.11	88	1.004542	62.1059	0.0161015	56.03984	3480.4	0.7121	–13.988
31.66	89	1.004717	62.0951	0.0161043	57.04177	3542.1	0.7351	–13.965

Temperature. Centig.	Temperature. Fahr.	Volume. Wat. = 1 at 40°.	Weight. per cubic foot.	Bulk. cub. per lb.	Units of heat. per lb.	Units of heat. pr. c. ft.	Pressure of vapor. Absol.	Pressure of vapor. under at.
t	𝔗	𝔙	𝔚	𝔈	*h.*	*h'.*	+ P.	— p.
32.22	90	1.004894	62.0840	0.016107	58.0437	3603.8	0.7586	—13.94
32.77	91	1.005094	62.0718	0.016110	59.0458	3665.0	0.7827	—13.91
33.33	92	1.005258	62.0617	0.016113	60.0479	3726.6	0.8075	—13.89
33.88	93	1.005444	62.0502	0.016116	61.0501	3788.2	0.8329	—13.86
34.44	94	1.005633	62.0386	0.016119	62.0523	3849.8	0.8590	—13.84
35.00	95	1.005825	62.0267	0.016122	63.0546	3911.2	0.8858	—13.81
35.55	96	1.006019	62.0148	0.016125	64.0569	3972.6	0.9132	—13.79
36.11	97	1.006216	62.0026	0.016128	65.0593	4033.9	0.9609	—13.74
36.66	98	1.006415	61.9904	0.016131	66.0618	4095.2	0.9704	—13.73
37.22	99	1.006618	61.9779	0.016135	67.0643	4156.5	1.000	—13.70
37.77	100	1.006822	61.9653	0.016138	68.0669	4217.7	1.030	—13.67
38.33	101	1.007030	61.9525	0.016141	69.0696	4278.9	1.061	—13.64
38.88	102	1.007240	61.9396	0.016145	70.0723	4340.1	1.093	—13.61
39.44	103	1.007553	61.9204	0.016150	71.0751	4401.3	1.126	—13.57
40.00	104	1.007668	61.9133	0.016152	72.0779	4462.5	1.159	—13.54
40.55	105	1.007905	61.8987	0.016155	73.0809	4523.0	1.194	—13.50
41.11	106	1.008106	61.8864	0.016159	74.0838	4585.0	1.229	—13.47
41.66	107	1.008328	61.8728	0.016162	75.0869	4645.9	1.265	—13.43
42.22	108	1.008554	61.8589	0.016166	76.0900	4706.8	1.302	—13.40
42.77	109	1.008781	61.8450	0.016169	77.0932	4767.7	1.340	—13.36
43.33	110	1.009032	61.8296	0.016173	78.0965	4828.6	1.378	—13.32
43.88	111	1.009244	61.8166	0.016177	79.0998	4889.5	1.418	—13.28
44.44	112	1.009479	61.8022	0.016180	80.1032	4950.4	1.459	—13.24
45.00	113	1.009718	61.7876	0.016184	81.1067	5011.3	1.500	—13.20
45.55	114	1.009956	61.7730	0.016188	82.1103	5072.2	1.543	—13.16
46.11	115	1.010197	61.7583	0.016192	83.1139	5133.0	1.587	—13.11
46.66	116	1.010442	61.7433	0.016196	84.1176	5193.7	1.631	—13.07
47.22	117	1.010688	61.7283	0.016200	85.1214	5254.3	1.677	—13.02
47.77	118	1.010938	61.7130	0.016204	86.1252	5314.9	1.723	—12.98
48.33	119	1.011189	61.6977	0.016208	87.1292	5375.5	1.771	—12.93
48.88	120	1.011442	61.6823	0.016212	88.1332	5436.1	1.820	—12.88
49.44	121	1.011698	61.6666	0.016216	89.1373	5496.6	1.870	—12.83
50.00	122	1.011956	61.6509	0.016220	90.1414	5557.1	1.921	—12.78
50.55	123	1.012216	61.6351	0.016224	91.1456	5617.6	1.974	—12.73
51.11	124	1.012478	61.6192	0.016229	92.1500	5678.1	2.026	—12.67
51.66	125	1.012743	61.6030	0.016233	93.1543	5738.6	2.082	—12.62
52.22	126	1.013010	61.5868	0.016237	94.1588	5798.9	2.137	—12.56
52.77	127	1.013278	61.5805	0.016241	95.1634	5859.2	2.195	—12.50
53.33	128	1.013550	61.5540	0.016246	96.1680	5919.5	2.253	—12.45
53.88	129	1.013823	61.5374	0.016250	97.1727	5979.7	2.312	—12.39
54.44	130	1.014098	61.5207	0.016255	98.1775	6040.0	2.374	—12.33
57.22	135	1.015505	61.4355	0.016277	103.2027	6340.3	2.699	—12.00
60.00	140	1.016962	61.3473	0.016301	108.230	6639.6	3.058	—11.64
62.77	145	1.018468	61.2567	0.016325	113.260	6937.9	3.462	—11.24
65.55	150	1.020021	61.1635	0.016350	118.291	7215.1	3.907	—10.79
68.33	155	1.021619	61.0678	0.016375	123.326	7531.2	4.397	—10.30
71.11	160	1.023262	60.9697	0.016401	128.362	7826.2	4.939	—9.761
73.88	165	1.024947	60.8695	0.016429	133.401	8098.1	5.534	—9.166
76.66	170	1.026672	60.7673	0.016456	138.443	8412.8	6.188	—8.512
79.44	175	1.028438	60.6620	0.016485	143.487	8704.2	6.906	—7.794
82.22	180	1.030242	60.5567	0.016513	148.537	8994.	7.693	—7.007
85.00	185	1.032083	60.4487	0.016543	153.583	9281.	8.550	—6.150
87.77	190	1.033960	60.3389	0.016573	158.635	9571.	9.488	—5.212
90.55	195	1.035873	60.2275	0.016604	163.691	9858.	10.51	—4.19
93.33	200	1.037819	60.1146	0.016635	168.749	10318.	11.62	—3.08
96.11	205	1.039798	60.0002	0.016667	173.809	10428.	12.83	—1.87
98.88	210	1.041809	59.8843	0.016799	178.873	10712.	14.13	—0.57
100.00	212	1.042622	59.8376	0.016811	180.900	18824.	14.70	0.000

TABLE XXXVIII.

Water.

Temperature of the water.		Volume. water = 1 at 40°.	Weight. lbs. per cubic ft.	Bulk. cubic feet per pound.	Units of heat in water from 32° to $\mathfrak{T}$.			
					Total per		Latent per	
Cent.	Fahr.				pound.	cubic foot.	pound.	cubic ft.
	$\mathfrak{T}$	$\mathfrak{V}$	$\mathfrak{W}$	$\mathfrak{E}$	$h.$	$h'.$	$l.$	$l'.$
100.	212.	1.04262	59.838	0.01671	180.90	10825	0.903	54.03
100.5	213.	1.04296	59.819	0.01671	181.91	10882	0.915	54.73
102.4	216.4	1.04436	59.743	0.01674	185.36	11063	0.957	56.73
104.2	219.6	1.04534	59.668	0.01676	188.59	11241	0.994	59.31
106.	222.8	1.04638	59.594	0.01678	191.83	11414	1.033	61.56
107.6	225.7	1.04785	59.520	0.01680	194.78	11583	1.082	64.40
109.1	228.5	1.04946	59.447	0.01682	197.63	11749	1.130	67.17
110.6	231.2	1.05062	59.384	0.01684	200.37	11895	1.170	69.48
112.1	233.8	1.05175	59.322	0.01685	203.01	12037	1.209	71.72
113.6	236.3	1.05284	59.261	0.01687	205.55	12175	1.248	73.96
114.8	238.7	1.05389	59.201	0.01689	207.98	12309	1.281	75.71
116.1	241.0	1.05490	59.142	0.01690	210.32	12439	1.322	78.19
117.7	243.3	1.05588	59.086	0.01692	212.66	12561	1.359	80.38
118.5	245.4	1.05683	59.032	0.01694	214.79	12678	1.394	82.42
119.7	247.5	1.05776	58.980	0.01695	216.84	12791	1.437	84.42
120.7	249.4	1.05867	58.930	0.01697	218.86	12901	1.462	86.32
121.8	251.4	1.05955	58.881	0.01698	220.90	13007	1.496	88.09
123.0	253.4	1.06042	58.832	0.01700	222.93	13113	1.532	90.02
124.0	255.3	1.06128	58.784	0.01701	224.86	13217	1.565	91.92
125.1	257.2	1.06213	58.737	0.01702	226.80	13318	1.598	93.78
126.1	259.0	1.06297	58.690	0.01704	228.63	13416	1.630	95.65
127.0	260.7	1.06380	58.646	0.01705	230.36	13510	1.664	97.59
128.0	262.4	1.06460	58.603	0.01706	232.09	13602	1.695	99.37
128.9	264.1	1.06538	58.561	0.01707	233.83	13692	1.726	101.1
129.8	265.7	1.06614	58.519	0.01709	235.45	13780	1.756	102.8
130.7	267.3	1.06689	58.477	0.01710	237.09	13866	1.790	104.5
131.6	268.9	1.06761	58.437	0.01711	238.72	13950	1.816	106.1
132.5	270.4	1.06832	58.398	0.01712	240.25	14036	1.846	107.9
133.4	271.9	1.06902	58.359	0.01713	241.78	14115	1.879	109.6
134.0	273.3	1.06971	58.321	0.01714	243.20	14192	1.905	111.2
134.9	274.8	1.07039	58.284	0.01716	244.73	14267	1.935	112.7
135.6	276.2	1.07105	58.250	0.01717	246.16	14339	1.961	114.2
136.4	277.6	1.07170	58.214	0.01718	247.59	14411	1.990	115.8
137.2	279.0	1.07234	58.179	0.01719	249.02	14482	2.018	117.4
137.9	280.3	1.07297	58.145	0.01720	250.34	14551	2.045	118.9
138.6	281.6	1.07359	58.112	0.01721	251.67	14620	2.075	120.3
139.3	282.8	1.07421	58.078	0.01722	252.90	14688	2.098	121.7
140.0	284.1	1.07483	58.045	0.01723	254.22	14755	2.126	123.2
140.8	285.4	1.07534	58.012	0.01724	255.66	14821	2.150	124.7
141.4	286.6	1.07594	57.980	0.01725	256.77	14886	2.175	126.2
142.0	287.8	1.07653	57.948	0.01726	258.00	14951	2.202	127.7

TABLE XXXIX.

Steam.

Total pressure.		Tem-perat're Fahr.	Volume water = 1 at 40.	Weight lbs. per cubic ft.	Bulk cubic ft. per lb.	Units of heat from 32° to T.				Pressure above atmosphere.
lbs. per sq. inch.	Inches mercur.					Total per		Latent per		
						pound.	cubic ft.	pound.	cubic ft.	
P	*I*	*T*	*V*	*W*	*C*	*H*	*H′*	*L*	*L′*	*p*
14.7	29.92	212	1740	0.0358	27.897	1146.6	41.100	965.7	34.61	.00
15	30.55	213	1706	0.0365	27.347	1147.0	41.920	965.1	35.29	.3
16	32.59	216.4	1601	0.0389	25.674	1148.0	44.700	962.7	37.50	1
17	34.63	219.6	1509	0.0413	24.186	1149.0	47.478	960.4	39.68	2
18	36.67	222.8	1426	0.0437	22.865	1149.9	50.255	958.1	41.86	3
19	38.71	225.7	1353	0.0461	21.693	1150.8	53.030	956.0	44.05	4
20	40.74	228.5	1288	0.0484	20.690	1151.7	55.802	954.1	46.23	5
21	42.78	231.2	1228	0.0508	19.678	1152.6	58.572	952.2	48.41	6
22	44.82	233.8	1173	0.0532	18.804	1153.4	61.340	950.7	50,48	7
23	46.85	236.3	1123	0.0555	18.005	1154.2	64.106	948.7	52.65	8
24	48.89	238.7	1078	0.0579	17.272	1155.0	66.870	946.0	54.82	9
25	50.93	241.0	1035	0.0602	16.597	1155.7	69.632	945.4	56.96	10
26	52.97	243.3	995.1	0.0625	15.994	1156.4	72.392	943.8	59.09	11
27	55.00	245.4	958.2	0.0648	15.422	1157.1	75.159	942.3	61.21	12
28	57.04	247.5	926.4	0.0672	14.881	1157.7	77.914	940.9	63.31	13
29	59.08	249.4	895.6	0.0696	14.371	1158.2	70.667	939.6	65.41	14
30	61.11	251.4	866.7	0.0720	13.892	1158.7	83.410	937.8	67.51	15
31	63.15	253.4	838.3	0.0743	13.456	1159.3	86.162	936.4	69.60	16
32	65.19	255.3	812.0	0.0766	13.059	1159.9	88.913	935.1	71.68	17
33	67.23	257.2	787.8	0.0789	12.669	1160.5	91.662	933.7	73.75	18
34	69.26	259.0	765.7	0.0812	12.313	1161.0	94.411	932.4	75.83	19
35	71.30	260.7	745.8	0.0834	11.955	1161.5	97.156	931.2	77.89	20
36	73.34	262.4	726.9	0.0860	11.624	1162.0	99.901	929.9	79.95	21
37	75.38	264.1	708.8	0.0884	11.309	1162.5	102.65	928.7	82.01	22
38	77.41	265.7	691.7	0.0908	11.013	1163.0	105.40	927.6	84.06	23
39	79.45	267.3	675.4	0.0930	10.745	1163.5	108.15	926.4	86.10	24
40	81.49	268.9	654.9	0.0952	10.498	1164.0	110.87	925.3	88.14	25
41	83.52	270.4	640.0	0.0974	10.262	1164.5	113.61	924.3	90.18	26
42	85.56	271.9	625.4	0.0997	10.031	1164.9	116.35	923.1	92.21	27
43	87.60	273.3	611.2	0.1020	9.8030	1165.4	119.09	922.1	94.24	28
44	89.64	274.8	597.4	0.1044	9.5801	1165.8	121.83	921.1	96.26	29
45	91.67	276.2	584.1	0.1068	9.3617	1166.2	124.57	920.1	98.28	30
46	93.71	277.6	571.9	0.1093	9.1465	1166.7	127.31	919.1	100.3	31
47	95.75	279.0	560.1	0.1117	8.9486	1167.2	130.05	918.0	102.3	32
48	97.78	280.3	548.8	0.1141	8.7596	1167.6	132.79	917.1	104.3	33
49	99.82	281.6	537.8	0.1166	8.5776	1168.0	135.53	916.2	106.3	34
50	101.86	282.8	527.2	0.1183	8.4504	1168.4	138.27	915.4	108.3	35
51	103.90	284.1	517.5	0.1206	8.2899	1168.8	141.00	914.5	110.3	36
52	105.93	285.4	507.1	0.1230	8.1284	1169.2	143.73	913.6	112.3	37
53	107.97	286.6	498.0	0.1254	7.9724	1169.5	146.46	912.7	114.3	38
54	110.01	287.8	489.2	0.1278	7.8249	1169.8	149.18	911.8	116.3	39

TABLE XL.

Water.

Temperature of the water. Cent.	Fahr.	Volume. water = 1 at 40°.	Weight. lbs. per cubic ft.	Bulk. cubic feet per pound.	Units of heat in water from 32° to T. Total per pound.	Total per cubic foot.	Latent per pound.	Latent per cubic ft.
	T	V	W	B	h.	h'.	l.	l'.
142.8	289.0	1.07720	57.917	0.01726	259.23	15014	2.230	129.2
143.4	290.2	1.07778	57.886	0.01727	260.46	15075	2.260	130.8
144.0	291.3	1.07835	57.857	0.01728	261.58	15135	2.286	132.2
144.6	292.4	1.07892	57.823	0.01729	262.71	15195	2.310	133.5
145.2	293.6	1.07943	57.795	0.01730	263.93	15254	2.335	134.7
145.9	294.7	1.07998	57.768	0.01731	265.05	15312	2.354	136.0
146.6	295.8	1.08051	57.739	0.01732	266.18	15368	2.382	137.4
147.1	296.9	1.08104	57.711	0.01733	267.30	15424	2.406	138.8
147.7	298.0	1.08157	57.683	0.01734	268.43	15480	2.430	140.2
148.3	299.0	1.08209	57.655	0.01735	269.45	15535	2.454	141.6
148.8	300.0	1.08259	57.629	0.01736	270.48	15588	2.480	142.9
149.3	301.0	1.08311	57.604	0.01737	271.50	15641	2.503	144.2
150.0	302.0	1.08362	57.579	0.01738	272.52	15693	2.525	145.5
150.5	303.0	1.08411	57.546	0.01738	273.55	15746	2.548	146.7
151.1	304.0	1.08460	57.522	0.01739	274.58	15797	2.572	147.8
151.6	305.0	1.08507	57.497	0.01740	275.60	15846	2.595	149.2
152.2	306.0	1.08556	57.472	0.01740	276.62	15896	2.618	150.4
152.8	307.0	1.08604	57.447	0.01741	277.64	15945	2.640	151.6
153.3	307.9	1.08653	57.420	0.01741	278.56	15995	2.658	152.8
153.8	308.9	1.08700	57.395	0.01742	279.58	16044	2.686	154.1
154.3	309.8	1.08747	57.370	0.01743	280.51	16093	2.707	155.3
154.8	310.7	1.08792	57.346	0.01743	281.43	16140	2.728	156.6
155.1	311.6	1.08838	57.322	0.01744	282.35	16187	2.755	157.9
155.9	312.5	1.08883	57.298	0.01745	283.27	16233	2.776	159.2
156.3	313.4	1.08928	57.275	0.01745	284.19	16278	2.795	160.4
156.8	314.3	1.08971	57.252	0.01746	285.12	16324	2.822	161.6
157.3	315.1	1.09014	57.230	0.01747	285.94	16368	2.840	162.7
157.7	315.9	1.09057	57.208	0.01747	286.76	16411	2.860	165.8
158.1	316.7	1.09100	57.186	0.01748	287.58	16453	2.881	164.8
158.6	317.5	1.09138	57.164	0.01749	288.40	16493	2.900	165.9
159.1	318.4	1.09180	57.142	0.01750	289.32	16533	2.920	166.9
159.6	319.2	1.09222	57.121	0.01750	290.14	16574	2.940	168.0
160.0	320.0	1.09264	57.100	0.01751	290.96	16614	2.960	169.1
160.4	320.8	1.09305	57.078	0.01752	291.78	16654	2.980	170.2
160.8	321.6	1.09346	57.057	0.01752	292.60	16695	3.000	171.3
161.2	322.4	1.09384	57.036	0.01753	293.42	16735	3.022	172.4
161.6	323.2	1.09425	57.015	0.01754	294.25	16774	3.047	173.5
162.2	324.0	1.09465	56.994	0.01754	295.07	16813	3.068	174.6
162.6	324.7	1.09506	56.973	0.01755	295.79	16852	3.089	175.7
163.0	325.4	1.09546	56.953	0.01755	296.5	16890	3.100	176.7

TABLE XLI.
Steam.

Total pressure. lbs. per sq. inch.	Total pressure. Inches mercur.	Temperat're Fahr.	Volume water = 1 at 40.	Weight lbs. per cubic ft.	Bulk cubic ft. per lb.	Units of heat from 32° to $\mathfrak{T}$. Total per pound.	Total per cubic ft.	Latent per pound.	Latent per cubic ft.	Pressure above atmosphere.
P	I	$\mathfrak{T}$	$\mathfrak{V}$	$\mathfrak{W}$	$\mathfrak{E}$	H	H'	L	L'	p
55	112.04	289.0	480.6	0.1298	7.7028	1170.1	151.91	910.9	118.3	40
56	114.08	290.2	472.1	0.1302	7.6774	1170.5	154.64	910.1	120.3	41
57	116.12	291.3	464.0	0.1324	7.5524	1170.9	157.37	909.9	122.2	42
58	118.16	292.4	456.2	0.1346	7.4277	1171.3	160.10	908.6	124.2	43
59	120.19	293.6	448.8	0.1388	7.2034	1171.6	162.83	907.7	126.1	44
60	122.23	294.7	441.6	0.1422	7.0786	1171.9	165.56	906.9	128.1	45
61	124.27	295.8	434.6	0.1434	6.9709	1172.3	168.28	906.1	130.0	46
62	126.30	296.9	427.8	0.1456	6.8643	1172.6	171.00	905.3	131.9	47
63	128.34	298.0	421.2	0.1479	6.7588	1172.9	173.71	904.5	133.9	48
64	130.38	299.0	414.9	0.1502	6.6543	1143.2	176.41	903.8	135.8	49
65	132.42	300.0	408.7	0.1526	6.5510	1173.5	179.13	903.0	137.8	50
66	134.45	301.0	402.6	0.1548	6.4570	1173.8	181.84	902.3	139.7	51
67	136.49	302.0	396.7	0.1571	6.3660	1174.1	184.53	901.6	141.7	52
68	138.53	303.0	391.1	0.1593	6.2750	1174.4	187.24	900.9	143.6	53
69	140.36	304.0	385.6	0.1616	6.1852	1174.7	190.00	900.1	145.6	54
70	142.60	305.0	380.4	0.1640	6.0972	1175.0	192.71	899.4	147.5	55
71	144.64	306.0	374.7	0.1662	6.0162	1175.3	195.42	898.7	149.5	56
72	146.68	307.0	369.5	0.1684	5.9363	1175.6	198.14	898.0	151.4	57
73	148.72	307.9	364.7	0.1707	5.8576	1175.9	200.85	897.4	153.3	58
74	150.75	308.9	360.2	0.1730	5.7799	1176.2	203.58	896.6	155.2	59
75	152.79	309.8	355.8	0.1753	5.7033	1176.5	206.29	896.0	157.1	60
76	154.83	310.7	351.1	0.1775	5.6324	1176.8	209.00	895.4	159.0	61
77	156.86	311.6	346.6	0.1798	5.5624	1177.1	211.71	895.8	160.9	62
78	158.90	312.5	342.3	0.1820	5.4933	1177.4	214.42	894.1	162.8	63
79	160.94	313.4	338.1	0.1843	5.4251	1177.6	217.13	893.4	164.7	64
80	162.98	314.3	334.3	0.1866	5.3576	1177.8	219.84	892.7	166.6	65
81	165.01	315.1	330.3	0.1888	5.2947	1178.1	222.55	892.2	168.5	66
82	167.05	315.9	326.4	0.1911	5.2327	1178.4	225.25	891.7	170.4	67
83	169.09	316.7	322.6	0.1926	5.1916	1178.7	227.96	891.1	172.3	68
84	171.12	317.5	318.8	0.1956	5.1114	1178.9	230.68	890.5	174.2	69
85	173.16	318.4	315.2	0.1979	5.0522	1179.1	233.38	889.8	176.1	70
86	175.20	319.2	311.7	0.2002	4.9955	1179.4	236.09	889.3	178.0	71
87	177.24	320.0	308.2	0.2024	4.9399	1179.7	238.79	888.8	179.9	72
88	179.27	320.8	304.8	0.2047	4.8855	1179.9	241.50	888.1	181.8	73
89	181.31	321.6	301.5	0.2069	4.8322	1180.1	244.21	887.5	183.6	74
90	183.35	322.4	298.2	0.2092	4.7803	1180.3	246.91	886.9	185.4	75
91	185.38	323.2	295.0	0.2114	4.7293	1180.6	249.62	886.4	187.3	76
92	187.32	324.0	291.9	0.2137	4.6794	1180.9	252.33	885.9	189.2	77
93	189.46	324.7	288.9	0.2159	4.6305	1181.1	255.04	885.3	191.0	78
94	191.50	325.4	285.9	0.2182	4.5827	1181.3	257.75	884.8	193.2	79

TABLE XLII.

Water.

Temperature of the water. Cent.	Fahr.	Volume. water = 1 at 40°.	Weight. lbs. per cubic ft.	Bulk. cubic feet per pound.	Units of heat in water from 32° to 𝔗. Total per pound.	cubic foot.	Latent per pound.	cubic ft.
𝔗	𝔗	𝔙	𝔚	𝔈	$h.$	$h'.$	$l.$	$l'.$
163.4	326.2	1.09578	56.934	0.01756	297.32	16928	3.121	177.7
163.8	327.0	1.09617	56.914	0.01756	298.14	16966	3.142	178.8
164.2	327.7	1.09655	56.894	0.01757	298.86	17004	3.163	179.9
164.6	328.5	1.09692	56.875	0.01758	299.68	17046	3.183	181.0
165.0	329.2	1.09730	56.855	0.01758	300.40	17078	3.204	182.1
165.4	329.9	1.09768	56.836	0.01759	301.12	17114	3.222	183.1
165.9	330.7	1.09804	56.818	0.01759	301.94	17149	3.240	184.1
166.3	331.3	1.09840	56.804	0.01760	302.56	17183	3.258	185.1
166.7	331.9	1.09876	56.786	0.01760	303.17	17217	3.276	186.0
167.0	332.6	1.09911	56.769	0.01761	303.89	17251	3.294	186.9
167.3	333.3	1.09949	56.743	0.01761	304.61	17284	3.312	187.9
167.7	334.0	1.09984	56.725	0.01762	305.33	17318	3.330	189.0
168.0	334.7	1.10019	56.706	0.01763	306.05	17350	3.349	190.0
168.4	335.4	1.10055	56.688	0.01763	306.77	17384	3.368	191.0
168.8	336.1	1.10091	56.670	0.01764	307.49	17427	3.387	192.0
169.2	336.8	1.10125	56.652	0.01764	308.21	17461	3.406	193.0
169.6	337.4	1.10159	56.635	0.01765	308.82	17493	3.425	194.0
170.0	338.0	1.10193	56.618	0.01766	309.44	17525	3.444	195.0
170.4	338.7	1.10226	56.600	0.01766	310.16	17557	3.462	196.0
170.8	339.4	1.10260	56.583	0.01767	310.88	17589	3.481	197.0
171.1	340.0	1.10292	56.566	0.01768	311.50	17621	3.500	198.0
172.9	343.2	1.10459	56.483	0.01770	314.79	17772	3.590	202.8
174.5	346.2	1.10627	56.403	0.01773	317.88	17921	3.678	207.5
176.2	349.2	1.10787	56.326	0.01775	320.96	18068	3.763	212.1
177.7	352.0	1.10940	56.236	0.01778	323.85	18212	3.850	216.5
179.2	354.8	1.11070	56.166	0.01780	326.73	18349	3.927	220.8
180.7	357.4	1.11208	56.098	0.01782	329.41	18481	4.010	225.0
182.2	360.0	1.11344	56.031	0.01784	332.09	18607	4.090	229.0
183.7	362.5	1.11478	55.965	0.01787	334.67	18730	4.168	233.3
185.0	365.0	1.11613	55.900	0.01789	337.24	18850	4.244	237.2
186.5	367.4	1.11742	55.834	0.01791	339.72	18966	4.318	241.0
188.0	369.8	1.11869	55.770	0.01793	342.19	19080	4.390	244.6
188.5	372.0	1.11993	55.708	0.01795	344.46	19190	4.460	248.5
190.0	374.2	1 12109	55.648	0.01797	346.73	19296	4.530	252.1
191.2	376.4	1.12227	55.591	0.01799	349.00	19399	4.598	255.7
192.5	378.5	1.12343	55.534	0.01800	351.16	19501	4.666	259.1
193.7	380.6	1.12456	55.477	0.01802	353.33	19602	4.731	262.5
194.4	382.6	1.12561	55.426	0.01804	355.39	19698	4.794	265.7
197.0	386.6	1.12783	55.317	0.01807	359.54	19885	4.940	272.8
199.1	390.4	1.13000	55.211	0.01811	363.48	20068	5.082	279.8

TABLE XLIII.
Steam.

Total pressure. lbs. per sq. inch.	Total pressure. Inches mercur.	Temperat're Fahr.	Volume water = 1 at 40.	Weight lbs. per cubic ft.	Bulk cubic ft. per lb.	Units of heat from 32° to T. Total per pound.	Total per cubic ft.	Latent per pound.	Latent per cubic ft.	Pressure above atmosphere.
P	I	T	V	W	C	H	H'	L	L'	p
95	193.53	326.2	283.0	0.2204	4.5361	1181.5	260.46	884.2	194.9	80
96	195.57	327.0	280.2	0.2227	4.4902	1181.8	263.16	883.8	196.7	81
97	197.61	327.7	277.4	0.2249	4.4454	1182.1	265.86	883.3	198.6	82
98	199.65	328.5	274.7	0.2271	4.4017	1182.3	268.55	882.6	200.4	83
99	201.68	329.2	272.0	0.2294	4.3591	1182.5	271.23	882.1	202.3	84
100	203.72	329.9	269.4	0.2316	4.3176	1182.7	273.93	881.6	204.2	85
101	205.76	330.7	266.8	0.2338	4.2769	1182.9	276.63	881.0	206.1	86
102	207.79	331.3	264.3	0.2360	4.2367	1183.1	279.32	880.6	208.0	87
103	209.83	331.9	261.8	0.2382	4.1970	1183.3	282.62	880.1	209.8	88
104	211.87	332.6	259.4	0.2405	4.1577	1183.5	284.70	879.6	211.6	89
105	213.91	333.3	257.0	0.2428	4.1187	1183.7	287.40	879.1	213.4	90
106	215.94	334.0	254.6	0.2450	4.0813	1183.9	290.09	879.6	215.2	91
107	217.98	334.7	252.3	0.2472	4.0444	1184.1	292.78	878.1	217.0	92
108	220.02	335.4	250.1	0.2495	4.0081	1184.3	295.48	877.5	218.9	93
109	222.06	336.1	247.9	0.2517	3.9723	1184.5	298.18	877.0	220.7	94
110	224.10	336.8	245.7	0.2540	3.9376	1184.7	300.87	876.5	222.6	95
111	226.13	337.4	243.5	0.2561	3.9036	1184.9	303.56	876.1	224.4	96
112	228.17	338.0	241.4	0.2584	3.8701	1185.1	306.26	875.7	226.3	97
113	230.20	338.7	239.3	0.2603	3.8411	1185.3	308.94	875.1	228.1	98
114	232.24	339.4	237.3	0.2628	3.8047	1185.5	311.65	874.6	229.9	99
115	234.28	340.0	235.3	0.2651	3.7722	1185.7	314.33	874.2	231.8	100
120	244.4	343.2	226.0	0.2759	3.6244	1186.6	327.89	873.8	241.0	105
125	254.6	346.2	217.2	0.2867	3.4875	1187.5	341.44	869.6	250.1	110
130	264.8	349.2	209.1	0.2984	3.3516	1188.4	355.00	867.4	259.0	115
135	275.0	352.0	201.4	0.3098	3.2278	1189.3	368.55	865.5	268.1	120
140	285.2	354.8	194.3	0.3212	3.1139	1190.1	381.88	863.5	277.0	125
145	295.4	357.4	187.8	0.3322	3.0105	1190.9	395.16	861.5	275.8	130
150	305.6	360.0	181.8	0.3432	2.9136	1191.7	408.38	859.6	294.5	135
155	310.8	362.5	176.5	0.3534	2.8289	1192.5	421.54	857.8	303.2	140
160	325.9	365.0	171.5	0.3646	2.7432	1193.3	435.08	856.1	312.1	145
165	336.0	367.4	166.6	0.3756	2.6617	1194.0	448.64	854.3	321.0	150
170	346.3	369.8	161.1	0.3871	2.5831	1194.7	462.22	852.5	329.9	155
175	356.5	372.0	157.0	0.3973	2.5171	1195.4	475.80	851.0	338.7	160
180	366.7	374.2	152.8	0.4075	2.4541	1196.1	488.96	849.4	347.1	165
185	376.9	376.4	148.8	0.4182	2.3916	1196.8	502.10	847.8	355.5	170
190	378.1	378.5	145.0	0.4292	2.3299	1197.4	515.20	846.2	363.9	175
195	387.3	380.6	141.5	0.4409	2.2684	1198.1	528.27	844.8	372.4	180
200	407.4	382.6	138.1	0.4517	2.2137	1198.7	542.07	843.3	381.0	185
210	427.8	386.6	132.0	0.4719	2.1192	1199.8	568.40	840.3	398.0	195
220	448.2	390.4	126.3	0.4935	2.0265	1201.0	574.70	837.5	414.8	205

TABLE XLIV.

Water.

Temperature of the water.		Volume. water = 1 at 40°.	Weight. lbs. per cubic ft.	Bulk. cubic feet per pound.	Units of heat in water from 32° to T. Total per		Latent per	
Cent.	Fahr.				pound.	cubic foot.	pound.	cubic ft.
T	T	V	W	E	h.	h'.	l.	l'.
201.1	394.0	1.13210	55.108	0.01814	367.20	20236	5.200	286.6
203.5	397.6	1.13301	55.017	0.01817	370.92	20402	5.318	292.9
205.0	401.0	1.13577	54.926	0.01821	374.44	20561	5.437	299.1
206.8	404.3	1.13760	54.838	0.01824	357.86	20720	5.558	305.2
208.7	407.5	1.13944	54.752	0.01826	381.18	20870	5.679	311.2
210.2	410.6	1.14119	54.670	0.01829	384.40	21015	5.800	317.1
211.9	413.5	1.14285	54.590	0.01832	387.40	21147	5.903	324.6
213.6	416.5	1.14441	54.514	0.01834	390.50	21273	6.006	332.0
215.1	419.2	1:14589	54.440	0.01837	393.31	21394	6.109	339.5
216.7	422.1	1.14743	54.367	0.01839	396.31	21510	6.212	346.7
218.2	424.8	1.14897	54.299	0.01841	399.11	21622	6.315	353.8
219.6	427.4	1.15050	54.230	0.01844	401.82	21751	6.418	356.9
221.1	430.0	1.15202	54.161	0.01846	404.52	21876	6.521	359.9
222.4	432.4	1.15339	54.093	0.01849	407.02	21997	6.624	362.8
223.6	434.9	1.15481	54.024	0.01851	409.63	22114	6.727	365.6
225.1	437.3	1.15621	53.959	0.01853	412.13	22238	6.830	368.5
226.4	439.6	1.15764	53.895	0.01856	414.53	22347	6.926	373.2
227.7	441.9	1.15880	53.834	0.01858	416.92	22452	7.020	377.9
228.9	444.1	1.16003	53.777	0.01859	419.21	22553	7.111	382.5
230.2	446.4	1.16127	53.721	0.01861	421.60	22650	7.200	386.9
231.4	448.5	1.16250	53.667	0.01863	423.79	22744	7.288	391.1
232.5	450.6	1.16372	53.614	0.01865	425.97	22843	7.374	395.3
233.6	452.6	1.16494	53.563	0.01867	428.06	22938	7.459	399.4
234.7	454.6	1.16571	53.513	0.01869	430.14	23029	7.542	403.6
235.9	456.7	1.16695	53.455	0.01871	432.32	23116	7.623	407.3
237.0	458.7	1.16818	53.406	0.01872	434.40	23200	7.700	411.2
238.0	460.6	1.16942	53.352	0.01874	436.38	23282	7.787	415.5
239.0	462.5	1.17066	53.293	0.01876	438.39	23363	7.893	423.3
241.1	466.1	1.17274	53.158	0.01881	442.21	23555	8.113	433.2
244.1	471.5	1.17598	53.027	0.01886	447.83	23741	8.329	442.9
246.5	475.7	1.17917	52.900	0.01890	452.24	23923	8.541	452.4
248.8	479.8	1.18231	52.768	0.01895	456.55	24091	8.747	461.6
253.1	487.6	1.18531	52.588	0.01901	464.66	24436	9.060	476.5
257.2	494.9	1.18961	52.430	0.01907	472.28	24762	9.381	491.8
261.0	501.8	1.19343	52.264	0.01913	479.51	25061	9.710	507.5
263.5	508.4	1.19742	52.102	0.01919	486.40	25577	10.00	521.0
268.1	514.6	1.20131	51.943	0.01925	492.97	25606	10.37	538.7
271.9	521.4	1.20562	51.787	0.01931	500.14	25901	10.74	556.2
273.3	526.0	1.20812	51.642	0.01936	505.00	26079	11.00	568.1
277.5	531.6	1.21147	51.498	0.01942	510.84	26307	11.242	578.8

TABLE XLV.

Steam.

Total pressure. lbs. per sq. inch.	Total pressure. Inches mercur.	Tem-perat're Fahr.	Volume water = 1 at 40.	Weight lbs. per cubic ft.	Bulk cubic ft. per lb.	Units of heat from 32° to 𝔗. Total per pound.	Units of heat from 32° to 𝔗. Total per cubic ft.	Units of heat from 32° to 𝔗. Latent per pound.	Units of heat from 32° to 𝔗. Latent per cubic ft.	Pressure above at-mosphere.
P	I	𝔗	𝔙	𝔚	𝔈	H	H'	L	L'	p
230	468.5	394.0	120.8	0.5165	1.9360	1202.2	620.96	835.0	431.3	215
240	488.9	397.6	116.1	0.5364	1.8646	1203.2	647.41	832.3	447.9	225
250	509.3	401.0	111.7	0.5595	1.7874	1204.2	673.85	829.8	464.4	235
260	529.7	404.3	107.5	0.4803	1.7230	1205.2	700.28	827.4	480.8	245
270	550.0	407.5	103.7	0.6016	1.6621	1206.2	726.66	825.0	497.1	255
280	570.4	410.6	100.2	0.6238	1.6031	1207.2	753.04	822.8	513.3	265
290	590.8	413.5	97.01	0.6459	1.5481	1208.1	779.40	820.7	529.4	275
300	611.1	416.5	94.22	0.6681	1.4967	1209.0	805.74	818.6	545.4	285
310	631.5	419.2	91.13	0.6896	1.4499	1209.8	832.96	816.5	561.4	295
320	651.9	422.1	88.21	0.7107	1.4071	1210.6	858.36	814.4	577.3	305
330	672.3	424.8	85.44	0.7302	1.3695	1211.5	884.63	812.4	593.2	315
340	692.6	427.4	83.19	0.7547	1.3250	1212.3	910.89	810.5	608.9	325
350	713.0	430.0	80.99	0.7745	1.2915	1213.1	937.13	808.6	624.5	335
360	733.4	432.4	78.84	0.7943	1.2590	1213.9	963.34	806.9	640.2	345
370	753.8	434.9	76.74	0.8146	1.2275	1214.7	989.51	805.1	655.8	355
380	774.1	437.3	74.66	0.8353	1.1968	1215.5	1015.7	803.4	671.3	365
390	794.5	439.6	72.90	0.8626	1.1597	1216.2	1041.8	801.7	686.7	375
400	814.9	441.9	71.19	0.8745	1.1434	1217.9	1067.9	800.0	702.0	385
410	835.2	444.1	69.52	0.8952	1.1170	1218.6	1094.0	799.4	717.2	395
420	855.6	446.4	67.90	0.9142	1.0938	1219.3	1120.2	797.7	732.4	405
430	876.0	448.5	66.34	0.9400	1.0634	1218.8	1146.3	795.0	747.6	415
440	896.4	450.6	64.91	0.9599	1.0417	1219.5	1172.3	793.5	762.8	425
450	916.7	452.6	63.55	0.9804	1.0201	1220.1	1198.3	792.0	777.9	435
460	937.1	454.6	62.22	1.0007	0.9993	1220.7	1124.3	790.5	792.9	445
470	957.5	456.7	60.94	1.0211	0.9793	1221.3	1150.4	789.0	807.8	455
480	977.8	458.7	59.72	1.0446	0.9573	1221.9	1276.5	787.5	822.7	465
490	998.2	460.6	58.54	1.0652	0.9388	1222.5	1302.3	786.1	837.4	475
500	1018.6	462.5	57.45	1.0859	0.9209	1223.0	1328.1	784.7	852.1	485
525	1069.5	466.1	54.81	1.1381	0.8786	1224.5	1392.6	782.3	881.8	510
550	1120.4	471.5	52.47	1.1890	0.8410	1225.8	1456.9	778.0	921.3	535
575	1171.4	475.7	50.32	1.2397	0.8066	1227.2	1521.0	775.0	960.4	560
600	1222.3	479.8	48.35	1.2901	0.7751	1228.3	1584.8	771.8	1000	585
650	1324.2	487.6	44.75	1.3943	0.7172	1230.6	1709.5	766.0	1082	635
700	1426.0	494.9	41.70	1.4961	0.6684	1232.7	1933.8	760.4	1157	685
750	1527.9	501.8	39.05	1.5977	0.6259	1234.9	2057.7	755.4	1234	735
800	1629.8	508.4	36.73	1.6986	0.5887	1237.0	2101.2	750.6	1307	785
850	1731.6	514.6	34.68	1.7989	0.5554	1238.9	2228.3	745.9	1374	835
900	1833.5	521.4	32.87	1.8979	0.5269	1241.0	2355.4	740.0	1435	885
950	1935.5	526.0	31.21	1.9992	0.5002	1242.4	2482.5	737.4	1490	935
1000	2037.2	531.6	29.73	2.0986	0.4765	1243.5	2609.6	732.3	1538	985

TABLE XLVI.

Mean Pressure of Expanding Steam.

Absolute steam pressure.	Grade of expansion of steam, denoted by X.							
	1.333	1.5	1.6	2	2.666	3	4	8
	Steam cut off at *l*, from beginning of stroke.							
P	$\frac{3}{4}$	$\frac{2}{3}$	$\frac{5}{8}$	$\frac{1}{2}$	$\frac{3}{8}$	$\frac{1}{3}$	$\frac{1}{4}$	$\frac{1}{8}$
0.5	0.4826	0.4683	0.4587	0.4232	0.3713	0.3497	0.2982	0.1924
1	0.9652	0.9367	0.9175	0.8465	0.7426	0.6995	0.5965	0.3849
2	1.9304	1.8734	1.8350	1.6931	1.4482	1.3991	1.1931	0.7698
3	2.8956	2.8100	2.7524	2.5396	2.2280	2.0986	1.7897	1.1548
4	3.8608	3.7468	3.6700	3.3862	2.8964	2.7982	2.3862	1.5396
5	4.8262	4.6835	4.5875	4.2328	3.7133	3.4977	2.9828	1.9246
6	5.7914	5.6202	5.5050	5.0794	4.4559	4.1972	3.5794	2.3095
7	6.7566	6.5569	6.4225	5.9260	5.1966	4.8967	4.1760	2.6944
8	7.7216	7.4936	7.3400	6.7726	5.9413	5.5963	4.7726	3.0794
9	8.6866	8.5303	8.2574	7.6192	6.6840	6.2958	5.3692	3.4643
10	9.6524	9.3670	9.1750	8.4657	7.4267	6.9954	5.9657	3.8493
11	10.617	10.304	10.092	9.3123	8.1694	7.6949	6.5622	4.2342
12	11.583	11.240	11.010	10.159	8.9121	8.3944	7.1589	4.6191
13	12.548	12.177	11.927	11.005	9.6548	9.0940	7.7555	5.0041
14	13.513	13.113	12.845	11.852	10.397	9.7935	8.3520	5.3890
15	14.478	14.050	13.762	12.698	11.140	10.493	8.9485	5.7739
16	15.443	14.987	14.679	13.545	11.882	11.192	9.5451	6.1588
17	16.408	15.923	15.597	14.392	12.625	11.892	10.141	6.5437
18	17.373	16.860	16.514	15.238	13.368	12.591	10.738	6.9287
19	18.339	17.797	17.432	16.085	14.110	13.291	11.335	7.3136
20	19.304	18.734	18.350	16.931	14.853	13.991	11.931	7.6986
21	20.269	19.671	19.268	17.778	15.596	14.690	12.527	8.0835
22	21.234	20.508	20.185	18.625	16.339	15.390	13.124	8.4684
23	22.199	21.545	21.103	19.471	17.082	16.089	13.720	8.8534
24	23.165	22.481	22.020	20.318	17.823	16.789	14.317	9.2383
25	24.130	23.481	22.938	21.164	18.567	17.488	14.913	9.6232
26	25.096	24.355	23.855	22.011	19.318	18.188	15.511	10.008
27	26.061	25.291	24.773	22.857	20.052	18.887	16.107	10.393
28	27.026	26.228	25.690	23.704	20.795	19.587	16.704	10.778
29	27.991	27.165	26.607	24.551	21.538	20.287	17.300	11.162
30	28.956	28.100	27.524	25.396	22.280	20.986	17.897	11.548
31	29.920	29.036	28.440	26.244	23.022	21.684	18.493	11.932
32	30.886	29.974	29.358	27.090	23.764	22.384	19.090	12.317
33	31.852	30.910	30.276	27.936	24.508	23.084	19.687	12.702
34	32.816	31.846	31.194	28.784	25.250	23.784	20.282	13.087
35	33.782	32.784	32.110	29.630	25.992	24.484	20.880	13.472
36	34.746	33.720	33.028	30.476	26.736	25.182	21.476	13.857
37	35.712	34.656	33.946	31.322	27.478	25.882	22.072	14.242
38	36.678	35.594	34.864	32.170	28.220	26.582	22.670	14.627
39	37.642	36.530	35.780	33.016	28.964	27.282	23.266	15.012

TABLE XLVII.

Mean Pressure of Expanding Steam.

Absolute steam pressure.	Grade of expansion of steam, denoted by X.							
	1.333	1.5	1.6	2	2.666	3	4	8
	Steam cut off at *l*, from beginning of stroke.							
P	$\frac{3}{4}$	$\frac{2}{3}$	$\frac{5}{8}$	$\frac{1}{2}$	$\frac{3}{8}$	$\frac{1}{3}$	$\frac{1}{4}$	$\frac{1}{8}$
50	48.262	46.835	45.875	42.328	37.133	34.977	29.828	19.246
55	53.088	51.518	50.462	46.561	40.846	38.474	32.811	21.170
60	57.914	56.202	55.050	50.794	44.559	41.972	35.794	23.095
65	62.740	60.885	59.637	55.027	48.273	45.470	38.777	25.020
70	67.566	65.569	64.225	59.260	51.986	48.967	41.760	26.944
75	72.393	70.252	68.812	63.493	55.700	52.465	44.743	28.869
80	77.216	74.936	73.400	67.726	59.413	55.963	47.726	30.794
85	82.042	79.619	77.987	71.959	63.126	59.461	50.709	32.718
90	86.866	85.303	82.574	76.192	66.840	62.958	53.692	34.643
95	91.699	89.986	87.163	80.425	70.553	66.456	56.675	36.568
100	96.524	93.670	91.750	84.657	74.267	69.954	59.657	38.493
105	101.35	98.353	96.337	88.890	77.981	73.451	62.640	40.417
110	106.17	103.04	100.92	93.123	81.694	76.949	65.622	42.342
115	111.00	107.72	105.51	97.356	85.407	80.447	68.606	44.267
120	115.83	112.40	110.10	101.59	89.121	83.944	71.589	46.191
125	120.65	117.08	114.68	105.82	92.834	87.442	74.572	48.116
130	125.48	121.77	119.27	110.05	96.548	90.940	77.555	50.041
135	130.30	126.45	123.86	114.28	100.26	94.437	80.538	51.966
140	135.13	131.13	128.45	118.52	103.97	97.935	83.520	53.890
145	139.96	135.82	133.03	122.75	107.68	101.43	86.502	55.815
150	144.78	140.50	137.62	126.98	111.40	104.93	89.485	57.739
155	149.60	145.18	142.20	131.22	115.11	108.42	92.468	59.663
160	154.43	149.87	146.79	135.45	118.82	111.92	95.451	61.588
165	159.26	154.55	151.38	139.68	122.54	115.42	98.434	63.513
170	164.08	159.23	155.97	143.92	126.25	118.92	101.41	65.437
175	168.91	163.92	160.55	148.15	129.96	122.42	104.40	67.362
180	173.73	168.60	165.14	152.38	133.68	125.91	107.38	69.287
185	178.56	173.28	169.73	156.61	137.39	129.41	110.36	71.212
190	183.39	177.97	174.32	160.85	141.10	132.91	113.35	73.136
195	188.21	182.65	178.90	165.08	144.82	136.41	116.33	75.061
200	193.04	187.34	183.50	169.31	148.53	139.91	119.31	76.986
210	202.69	196.71	192.68	177.78	155.96	146.90	125.27	80.835
220	212.34	205.08	201.85	186.25	163.39	153.90	131.24	84.684
230	221.99	215.45	211.03	194.71	170.82	160.89	137.20	88.534
240	231.65	224.81	220.20	203.18	178.23	167.89	143.17	92.383
250	241.30	234.18	229.38	211.64	185.67	174.88	149.13	96.232
260	250.96	243.55	238.55	220.11	193.18	181.88	155.11	100.08
270	260.61	252.91	247.73	228.57	200.52	188.87	161.07	103.93
280	270.26	262.28	256.90	237.04	207.95	195.87	167.04	107.78
300	289.56	281.00	275.24	253.96	222.80	209.86	178.97	115.48

STRENGTH OF SPHERICAL SHELLS OF STEAM-BOILERS.—Addendum to § 86, page 105.

§ 126. For a spherical shell the tension or strain is equal to the area of the great circle in square inches multiplied by the steam pressure per square inch, which is resisted by the section of the shell in the great circumference.

When only a part of the sphere is used, like in spherical ends of boilers or steam-drums, the same rule holds good, only that the strength must be calculated for the whole sphere.

R = radius of the sphere in inches.
p = steam pressure in pounds per square inch.
t = thickness of shell in fraction of an inch.
S = ultimate strength of the iron in pounds per square inch.

Action of steam, $p \pi R^2 = S t 2 \pi R$, the reaction of the shell.

Ultimate Strength of Solid Shell in the Sphere without Riveted Joints.

Steam pressure, $p = \frac{2tS}{R}$. 1

Radius of sphere, $R = \frac{2tS}{p}$. 2

Thickness of shell, $t = \frac{Rp}{2S}$. 3

Breaking-strain, $S = \frac{Rp}{2t}$. 4

Example 1. The spherical end of a boiler is made of iron stamped $S = 60,000$ and $t = 0.25$ of an inch thick in one sheet without joints. What steam bursting-pressure can that spherical end stand with a radius of curvature $R = 96$ inches?

Steam-pressure, $p = \frac{2 \times 0.25 \times 60000}{96} = 312.5$ pounds.

These formulas are the same as those for cylindrical shells, with the exception that the radius R of the sphere takes the place for the diameter D of the cylinder. Therefore a sphere is double as strong as a cylinder of the same diameter. The coefficient X for safety strength will therefore be the same as for cylindrical shells, § 86, page 105, namely,

TABLE XXVI.

Coefficients X for Spherical Ends.

Construction of Shell.	X	Per cent. of strength.
Solid plate without joints....................	0.5	100
Double-riveted drilled holes................	0.4	80
Double-riveted punched holes.............	0.35	70
Single-riveted drilled holes.................	0.3	60
Single-riveted punched holes...............	0.25	50

Steam-pressure, $$p = \frac{XtS}{R}. \quad \ldots \quad 5$$

Radius of shell, $$R = \frac{XtS}{p}. \quad \ldots \quad 6$$

Thickness of plate, $$t = \frac{Rp}{XS}. \quad \ldots \quad 7$$

Breaking-strain, $$S = \frac{Rp}{Xt}. \quad \ldots \quad 8$$

The radius R, of the spherical end, is independent of the diameter D, of the boiler or steam-drum.

Example 6. What radius is required for a spherical boiler-end of solid plate $t = 0.3$ of an inch thick and stamped $S = 64{,}000$ to bear with safety a steam-pressure of $p = 80$ pounds per square inch?

Radius, $$R = \frac{0.5 \times 0.3 \times 64000}{80} = 120 \text{ inches.}$$

Example 7. The iron for a spherical boiler-end is expected to bear $S = 56{,}000$ pounds to the square inch of section, is to be curved to a radius $R = 84$ inches, and to have one double-riveted lap-joint with punched holes, and to bear a steam-pressure of $p = 96$ pounds to the square inch. Required the thickness of the iron?

Thickness, $$t = \frac{84 \times 96}{0.35 \times 56000} = 0.411 \text{ of an inch.}$$

PHYSICAL PROPERTIES OF DIFFERENT KINDS OF VAPORS.

§ 127. The following Table 48 shows the relation between temperature and pressure of vapors composed of the four principal simple elements—namely, *oxygen*, *nitrogen*, *hydrogen* and *carbon*. The table is deduced from the experiments of Regnault, except the column for carbonic acid, which is deduced from the experiments of Faraday and Pelouze; but those experimenters are not responsible for the formulas and tables which the writer has deduced from their experiments.

The vapors of *water* and *carbonic acid* have been treated in the preceding pages, and the next in order in the table is *turpentine*.

Oil of Turpentine is distilled from resin of pine trees. It is a volatile spirit composed of $C_{10}\,H_{16}$, and boils under atmospheric pressure at a temperature of 338° Fahr. The table gives the pressure under which it boils at different temperatures.

The formulas for pressure and temperatures of turpentine vapor are

$$T = 281\sqrt[6]{P} - 115.$$

$$P = \left(\frac{T+115}{281}\right)^6$$

Turpentine is a transparent liquid or gas insoluble in water, but dissolves paints and many gums and resins.

Alcohol.—Pure alcohol, $C_4H_6O_2$, boils under atmospheric pressure at a temperature of 173° Fahr. The formulas for pressure and temperature of alcoholic vapor are

$$T = 180\sqrt[6]{P} - 108. \qquad 1$$

$$P = \left(\frac{T+108}{180}\right)^6. \qquad 2$$

The ideal zero of vapor of alcohol, according to the formula, should be $-108°$ below Fahr. zero.

The pressure of vapor of alcohol is about double that of steam of equal temperature, as will be seen in the Table. The vapor of alcohol has been tried in France as motive power, and a large passenger steamer named "Kabyl," built in the year 1857, was supplied with engines and boilers for the use of alcohol instead of water. The

"Kabyl" was running from Marseilles to ports in the Mediterranean in the year 1858 with partial success, but the alcohol was finally abandoned for the reason that its saving in fuel did not compensate for the leakage of the more expensive fluid.

The vapor of the alcohol was condensed in an ordinary tubular fresh-water condenser and returned to the boiler, thus used over again perpetually.

The difficulty appeared to be the leakage of alcohol, and consequently the expense of supplying that fluid. The writer was on board the "Kabyl" during the first trial trip, but the memorandum then made has been lost. The first trial was made with *ether*, which was gradually converted into alcohol—that is, one atom of oxygen and one of hydrogen formed water—but even with this change in the fluid the consumption of fuel proved to be very economical.

One great advantage in using alcohol or ether instead of water in steam-boilers is that no incrustation is formed.

There was a very strong, but rather pleasant, odor of alcohol all over the ship, of which the passengers did not seem to complain.

Ether.—Pure ether, C_4H_5O, boils under atmospheric pressure at a temperature of 97° Fahr. The pressure of vapor of ether is five to six times that of steam of equal temperature, as seen in the accompanying table.

The formulas for pressure and temperature of etheric vapor are

$$T = 200\sqrt[6]{P} - 216. \qquad 1$$

$$P = \left(\frac{T+216}{200}\right)^6. \qquad 2$$

The ideal zero is $-216°$.

Benzine is a transparent liquid insoluble in water and dissolves fatty matter. It boils under atmospheric pressure at a temperature of 185° Fahr.

The following Table L. shows the boiling point of benzine under different pressures.

The formulas for pressure and temperature of vapor of benzine are

$$T = 222\sqrt[6]{P} - 162. \qquad 1$$

$$P = \left(\frac{T+162}{222}\right)^6. \qquad 2$$

Ammonia, NH_3, is a colorless vapor or liquid which boils under atmospheric pressure at about $-19.3°$ below Fahr. zero. The specific gravity of the liquid is about 0.76, and according to Faraday's experiments, freezes to a white transparent solid at $-103°$ Fahr., at which temperature the pressure of its vapor is about 5 pounds to the square inch. Ammonia is soluble in water, with which it generates heat, forming aqueous ammonia of great expansibility.

The high tension of ammonia at low temperatures is made use of in producing cold, for which purpose liquid ammonia is kept under very high pressure in a vessel, from which a small quantity is allowed to gradually escape into another vessel or tube, where it instantly evaporates, and the heat absorbed by that evaporation produces a very low temperature of the surrounding vessel or tube, so that water in the neighborhood will freeze to ice. This is the principle upon which ice-machines are constructed.

The formulas for pressure and temperature of vapor of ammonia are

$$T = 150\sqrt[6]{P} - 254. \quad . \quad . \quad . \quad . \quad 1$$

$$P = \left(\frac{T + 254}{150}\right)^6. \quad . \quad . \quad . \quad . \quad . \quad 2$$

Protoxide of Nitrogen, NO. This vapor is also called nitrous oxide or laughing gas, from its peculiar effect upon the mind when inhaled.

The specific gravity of nitrous oxide is 1.524.

The formulas for pressure and temperature of protoxide of nitrogen are

$$T = 175\sqrt[6]{P} - 464. \quad . \quad . \quad . \quad . \quad 1$$

$$P = \left(\frac{T + 464}{175}\right)^6. \quad . \quad . \quad . \quad . \quad . \quad 2$$

The last column in the table shows the pressure per square inch of nitrous oxide, corresponding to the temperatures in the first columns.

The Roman numbers in the table are converted from Regnault's experiments,* and the Italic numbers are calculated by the respective formulas.

The object in giving this table is to show at a glance the widely different physical properties of vapors composed of only oxygen, nitrogen, hydrogen and carbon.

* Memoires de l'Academie de France, Tome XXVI.

TABLE XLVIII.

Temperature and Pressure in Pounds per Square Inch of Different Kinds of Vapor.

Temperatures. Cent.	Temperatures. Fahr.	Water, Steam.	Carbonic acid.	Turpentine.	Alcohol.	Ether of alcohol.	Benzene.	Ammonia.	Protoxide of nitrogen.
T	T	HO	CO_2	$C_{10}H_{16}$	$C_4H_6O_2$	C_4H_5O	$C_{12}H_6$	NH_3	NO
—40	—40	....	*164.8*	....	....	*0.464*	....	8.4	*202*
—35	—31	....	*193.4*	....	....	*0.626*	....	12.0	*246*
—30	—22	0.007	*225.7*	....	....	*0.833*	*0.025*	16.72	*270*
—25	—13	0.012	*261.8*	....	....	*1.092*	*0.049*	21.4	304
—20	— 4	0.18	*302.1*	....	0.064	1.33	0.112	26.9	340
—15	+ 5	0.027	*346.9*	....	0.098	1.73	0.17	33.6	381
—10	14	0.040	*396.5*	....	0.125	2.22	0.25	41.6	425
— 5	23	0.060	*451.2*	....	0.176	2.82	0.355	50.8	476
0	32	0.089	*514.5*	0.04	0.245	3.57	0.489	61.6	530
+ 5	41	0.127	*577.4*	0.047	0.341	4.47	0.66	74.	591
10	50	0.177	*649.6*	0.057	0.469	5.54	0.875	88.4	658
15	59	0.246	*735.0*	0.069	0.638	6.84	1.14	105.	732
20	68	0.337	*814.2*	0.086	0.859	8.37	1.46	123.2	813
25	77	0.456	*886.6*	0.105	1.15	10.2	1.85	145	903
30	86	0.61	*1008*	0.133	1.517	12.27	2.33	168	1000
35	95	0.808	*1117*	0.151	2.	14.7	2.89	195	1110
40	104	1.06	*1234*	0.208	2.583	17.55	3.55	223.8	1225
45	113	1.38	*1362*	0.257	3.33	20.8	4.34	258	*1300*
50	122	1.78	*1471*	0.328	4.25	24.42	5.24	293	*1400*
55	131	2.27	*1644*	0.405	5.38	28.7	6.3	333	*1520*
60	140	2.88	*1817*	0.511	6.78	33.33	7.54	376.5	*1686*
65	149	3.61	*1968*	0.631	8.44	38.7	8.97	425	*1838*
70	158	4.51	*2147*	0.785	10.45	45.4	10.6	476	*2018*
75	167	5.58	*2352*	0.958	12.9	51.2	12.4	534	*2231*
80	176	6.86	*2542*	1.183	15.71	58.4	14.65	596	*2403*
85	185	8.37	*2758*	1.451	19.1	66.5	16.9	664	*2607*
90	194	10.2	*2988*	1.75	23.	75.3	19.6	736	*2825*
95	203	12.26	*3232*	2.123	27.6	77.4	22.6	816	*3082*
100	212	14.7	*3500*	2.54	32.8	95.8	26.	900	*3359*
105	221	17.5	*3770*	3.00	38.9	108	29.7	*1008*	*3627*
110	230	20.8	*4060*	3.59	45.75	120	33.7	*1135*	*3926*
115	239	24.5	*4369*	4.22	53.6	134	38.2	*1268*	*4220*
120	248	28.8	*4695*	4.76	62.6	149.3	43.2	*1425*	*4558*
125	257	33.8	*5026*	4.86	72.4	*165*	48.7	*1572*	*4926*
130	266	39.3	*5394*	6.73	83.6	*194*	54.6	*1745*	*5272*
135	275	45.5	*5769*	7.85	*96.*	*218*	61.	*1934*	*5727*
140	284	52.5	*6165*	8.97	*109.9*	*245*	66.	*2143*	*6087*
145	293	69.3	*6586*	10.35	*125.*	*270*	75.7	*2364*	*6590*
150	302	79.	*7015*	11.7	*141.5*	*300*	83.8	*2607*	*7061*
155	311	79.	*7470*	13.25	*159.8*	*335*	*88.1*	*2879*	*7556*
160	320	90.	*7984*	15.	*187.1*	*354*	*96.7*	*3156*	*8128*
165	329	102.	*8462*	16.9	*214.3*	*409*	*105.9*	*3481*	*8710*
170	338	115.5	*9000*	18.9	*245.6*	*451*	*115.8*	*3798*	*9253*
175	347	130.	*9552*	21.	*283.4*	*497*	*126.2*	*4157*	*9914*
180	356	146.	....	23.4	*320*	*547*	*137.4*	*4545*	
185	365	163.5	....	25.9	*360*	*601*	*149.3*	*4962*	
190	374	183.	....	28.5	*401*	*659*	*162.0*	*5411*	
195	383	203.5	....	31.3	*443*	*722*	*175.5*	*5892*	
200	392	226.	....	34.2	*490*	*789*	*189.*	*6444*	

§ 128. **BOILING POINT UNDER ATMOSPHERIC PRESSURE.**

$$\sqrt[6]{14.7} = 1.565.$$

Water,	$T = 200\sqrt[6]{14.7} - 101 = +212°.$
Carbonic acid,	$T = 61.404\sqrt[4]{14.7} - 260 = -140°.$
Turpentine,	$T = 281\sqrt[6]{14.7} - 115 = +324.7°.$
Alcohol,	$T = 180\sqrt[6]{14.7} - 108 = +173.7°.$
Ether,	$T = 200\sqrt[6]{14.7} - 216 = +97°.$
Benzine,	$T = 222\sqrt[6]{14.7} - 162 = +185.4°.$
Ammonia,	$T = 150\sqrt[6]{14.7} - 254 = -19.3°.$
Protoxide of nitrogen,	$T = 175\sqrt[6]{14.7} - 464 = -190.2°.$

BOILING POINT OR TEMPERATURE OF DISTILLATION OF PETROLEUM OILS.

§ 129. The variety of oils distilled from petroleum boil at widely different temperatures, according to their specific gravity. The boiling point under atmospheric pressure varies, as the cube of the specific gravity, from the ideal zero $-215°$ Fahr.

S = specific gravity of the oil compared with water as 1 at 32°.
T = temperature Fahr. at which the oil boils or distills under atmospheric pressure.

Boiling point, $T = 1150\, S^3 - 215°.$. 1

Specific gravity, $S = \sqrt[3]{\dfrac{T + 215°}{1150}}.$ . . . 2

Example 1. The specific gravity of Kerosene oil is 0.808. Required its boiling point?

$$T = 1150 \times 0.808^3 - 215° = 491.6°.$$

TEMPERATURE OF INFLAMMATION OF OILS DISTILLED FROM PETROLEUM.

§ 130. The volatility of distilled petroleum oils under atmospheric pressure ceases to exist under a certain temperature depending upon the sixth power of the specific gravity of the oil. Above that temperature the oil evaporates and mixes with the air, and can be ignited by a lighted match.

t = lowest temperature of inflammation, Fahr.
S = specific gravity of the oil, water = 1.

$$t = 1200\,S^6 - 140^\circ. \qquad . \quad . \quad . \quad . \quad 3$$

$$S = \sqrt[6]{\frac{t+140}{1200}} \qquad . \quad . \quad . \quad . \quad . \quad 4$$

Undistilled or mixed oils will ignite at a lower temperature than this formula. Crude petroleum ignites at 60°.

Example 3. Required the lowest temperature of inflammation of Kerosene oil of specific gravity 0.805?

$$t = 1200 \times 0.805^6 - 140 = 180^\circ.$$

TABLE L.

Temperatures of Distillation and Inflammation of Petroleum Oils.

Sp. gr. S	Names of Petroleum Oils.	Distillation. Fahr.	Distillation. Cent.	Inflammation. Fahr.	Inflammation. Cent.
0.6000	....	34°	1.11°	−84°	−65°
0.6125	....	49	9.44	−76	−60
0.625	Rhigolene...........................	63	17.22	−68	−55
0.6375	....	83	28.33	−59	−51
0.6500	Amylene...............................	101	38.33	−49	−45
0.6625	Gasolene..............................	119	48.33	−38	−39
0.675	....	139	59.44	−26	−32
0.6875	....	159	70.55	−13	−25
0.7000	Benzine................................	180	82.22	2	−16
0.7125	....	201	93.88	18	−7.7
0.7250	Toluene................................	219	103.8	35	+1.66
0.7375	Naphtha...............................	246	118.8	54	12.2
0.7500	Naphtha or Xylene..................	270	132.2	74	23.3
0.7625	Naphtha or Pyridine...............	295	146.1	97	36.1[5]
0.7750	Lutidine................................	320	160.0	121	49.4
0.7875	Aniline.................................	347	187.7	142	61.1
0.8000	Kerosene..............................	375	190.5	176	79.4
0.8125	Anthracene...........................	402	205.5	207	97.2
0.8250	Naphthaline...........................	424	217.7	240	115
0.8375	Paraffine..............................	460	237.7	276	135
0.850	Mineral Sperm Oil..................	490	254.4	314	156
0.8625	....	524	273.3	356	180
0.8750	....	555	290	399	204
0.8875	Lubricating Oil.......................	589	304.4	447	230
0.9000	....	623	328.3	498	259

APPENDIX.

TECHNICAL TERMS IN MECHANICS.

THE science of Mechanics has heretofore been afflicted with a language of vague terms promiscuously used without definite meaning, so that different ideas have been formed from one and the same expression and a variety of terms have been employed to express one and the same principle.

The most crucial test of perfection of a science is precision in its vocabulary and perspicuity in its principles, so that each expression bears a definite meaning.

The writer has for many years labored upon this subject—namely, to expel some indefinite terms and expressions which have heretofore embarrassed the science of Mechanics. In discussing the subject he has encountered difficulties with learned men, many of whom appear to have only faith in the old dogmas, and have thus thrown obstacles in the way of success.

Mr. William Dennison of East Cambridge, Mass., was the first one who understood and acknowledged the correctness of the new classification of dynamic elements and functions, and of their respective definitions. Mr. Dennison addressed the author on the subject as follows:

EAST CAMBRIDGE, MASS., May 12, 1874.

MR. JOHN W. NYSTROM,

Dear Sir—In reading your pamphlet on Dynamics I have been greatly interested, as I always am on all such subjects; but this subject should interest every one especially until its proper terms be adopted and their meaning permanently established. Except among mechanics you will seldom find any two persons to have the same ideas upon this subject, notwithstanding assertions to the contrary.

The very fact that the simple question of force of a falling body was discussed by so many learned men, all with different ideas on the subject, and no two of them agreed as to which is right, is sufficient proof of the present confusion in Dynamics.

Your reply to these jarring opinions, as well as to all other assertions in the pamphlet, is forcible, correct and to the purpose.

I consider the basis upon which you have placed this subject to be firm and well constructed, and of such a nature as never to be overthrown or destroyed.

You have also succeeded admirably in placing the subject in the most clear, comprehensive and proper light.

Had there been such a treatise in our schooldays, it would have been of the greatest assistance to us all, then and since. But this subject has always been in such a state of confounded conglomeration that we have been obliged to rely upon our own reasoning powers and practical understanding; therefore but few comparatively have been able to master the subject.

I have often been impressed with the idea that some scientific men like to flourish high-sounding terms, such as those you have rejected as useless and confusing. They often display extraordinary ability in compiling highly scientific terms into heaps of phrases which may appear learned to those not familiar with the subject, whilst they are sometimes mere inventions of words pretending to represent mysterious phenomena. Yours truly,

WILLIAM DENNISON.

In a pamphlet on dynamical terms the writer invited institutions of learning to discuss the subject, which invitation was accepted by many, of which a few sided with the writer; but the majority were against his views. The response of Professor Gustav Schmidt, of the Polytechnic Institute at Prague, in Bohemia, may serve as an average illustration of the present condition of the science of Mechanics in institutions of learning. The ideas on the subject held by others are substantially the same as those of Prof. Schmidt.

In the following pages, the comments of Prof. Schmidt are on the left-hand and the answers on the right-hand pages, so that the numbers of the paragraphs of the comments correspond to the numbers of the answering paragraphs.

The division into paragraphs has been made by the author.

(Translation from the German.)

MR. JOHN W. NYSTROM,

Dear Sir—It affords me great pleasure to comply with your request for a written opinion on your work, "Principles of Dynamics," and will do so in German on account of my insufficient knowledge of the English language.

§ 1. I have no objection to your answering me publicly in an American journal, provided you would publish an idiomatic translation of this letter.

§ 2. The term "Pferde-kraft" (horse-power) has become obsolete in Germany, and has been replaced by the term "Pferde-stärke" (horse-strength), as proposed by Reuleaux. The product $\mathfrak{P} = F\,V = \frac{K}{T}$ should consequently be called horse-strength.

§ 3. It is customary, however, to use the word "effect," but not the word "kraft" (force), as under no circumstance would it answer for the German idiom to use the term "kraft" (power) for "effect" or "pferde-stärke" (horse-strength or force).

§ 4. The former Prussian "pferde-stärke" undoubtedly had 513 second foot-pounds or 480 foot-pounds of the new weight; this, however, is not 582, but 544.8 English second foot-pounds.

§ 5. The present German "pferde-stärke" has, as in France, 75 second-metre kilogrammes = 542.5 English foot-pounds.

§ 6. The unit proposed by you—namely, 500 English foot-pounds—would be $69\frac{1}{8}$, or nearly 70 metre kilogrammes, equal to the performance of a horse at the plough.

§ 7. As, however, the English measurement will probably give way to that of the French during this century, the 75 *M. K.* already adopted will most probably be retained.

§ 8. The product $F\,T$ (dynamical moment, as you call it) is never used. It could have a meaning only if the force F remains constant during the time T; then most certainly for a uniformly accelerated motion from a state of rest, $F\,T$ would be $= M\,V$.

§ 9. However, for a uniformly accelerated motion with an initial velocity C, $F\,T = M(V - C)$; for instance, in the case of a vertical projection

$$F = -W, \qquad \text{then} \qquad W\,T = M(C - V) = \frac{W}{g}(C - V).$$

$$g\,T = C - V \qquad \text{and} \qquad V = C - g\,T.$$

PROFESSOR GUSTAV SCHMIDT,

Dear Sir—It affords me great pleasure to answer your comments on my "Principles of Dynamics," and I hope the translation of your paper from German to English is satisfactory to you.

§ 1. No American journal would publish this kind of discussion, for which reason I have concluded to append the same to this work on "Steam Engineering."

§ 2. Both the terms "kraft" and "stärke" in the German language mean "force." You have no German word for the function $\mathfrak{P} = F\ V$, which is power. Both your terms for horse-power mean horse-force. Strength or "stärke" is the capability of resisting static force.

The products
$$\begin{cases} \mathfrak{P} = F\ V = \dfrac{K}{T} \text{ is power in effects.} \\ \text{HP} = \dfrac{F\ V}{550} = \dfrac{K}{550\ T} \text{ is horse-power.} \end{cases}$$

The term "Pferde-kraft" is more proper than "Pferde-sterke."

§ 3. You say it is customary to use the word "effect" and give the other terms for which it is not used, but do not state for what it is used or what are its constituent elements. The term "effect" represents a unit of measurement of power—namely, a second foot-pound of power. Horse-power is another unit of power, consisting of 550 effects. You do not distinguish power from force in your language.

§ 4. According to the data of Prussian weight and measure in my possession—namely, 1.0297 ft. × 1.1023 lbs. × 513 = 582.18 English foot-pounds. This, however, does not affect the correctness of the principles of Dynamics.

§ 5. I gave 542.47 English second foot-pounds per 75 second-metre kilogrammes, and did not know the new Prussian measures.

§ 6. This unit was proposed only to accommodate the English weight and measure for the easy calculation and estimation of horse-power and practice.

§ 7. It is yet doubtful whether the English measurement will give way for that of the French in the present century, of which only 24 years remain.

§ 8. Because the momentum $F\ T$ is not used, is the reason why confusion still pervades the dynamics of matter. This momentum is there, whether it is used or not. When F is the mean force in the time T, the momentum must always be $F\ T = M\ V$.

§ 10. For a variable force F, however,

$$F\,\partial t = M\,\partial v, \text{ or}$$

$$F = M\frac{\partial v}{\partial t} = \frac{W}{g}\cdot\frac{\partial v}{\partial t}, \text{ as } g' = \frac{\partial v}{\partial t} = g\frac{F}{W}.$$

§ 11. Only this equation will answer for a general application; $F = \frac{M\ V}{T}$ (force of a moving body), on the contrary, is quite superfluous and inadmissible idea, as T, and consequently F, would be entirely arbitrary.

§ 12. You entirely omit the above-mentioned highly important term $g' = \frac{\partial v}{\partial t} = \frac{F}{M}$, which is the acceleration.

§ 13. For "work" in a moving body, $K = \frac{1}{2}M\ V^2 = W\frac{V}{g}$, the old term "lebendigo-kraft," living force, also sometimes "energie," energy, is used in Germany. I have proposed for it "bervegungs arbeit," work of motion, to distinguish it from "verschriebungs arbeit," work of pushing or drawing, $F\ S$ or universally $\int E\ \partial s$.

§ 14. We do not designate the value $\frac{1}{2}MV^2$ "Grösse der Bervegung," Quantitat der Bervegung (quantity of motion), but the product MV which you call (Bervegungs moment) moment of motion.

§ 15. You reject the term "acting force" and "working force." If, however, the mass M is moved by a force F, which is exactly equal to the sum of all resistances F', and its velocity V is consequently invariable, as, for instance, in the case with a train of cars, then F is a "working force" producing the pushing or pulling work $k = FS$, which is consumed by the equally great resistance $F'K' = F'S$. Therefore the force F cannot cause any acceleration of speed.

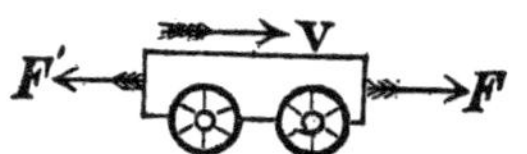

If the force F is greater than the resistance F', then there remains an accelerating force $f = F - F'$, which imparts to the mass $M = \frac{W}{g}$ the acceleration $g' = \frac{\triangle V}{\triangle T} = \frac{f}{M}$, if f is a constant quantity, or if f is invariable it imparts the acceleration $g' = \frac{\partial V}{\partial t} = \frac{f}{M}$. This accelerating force $f = F - F'$ must not be mistaken for a non-accelerating but "working force" F, nor for a non-working but only "*deformirender*"

§ 9. Motion and rest are only relative, for which reason the velocity V must always mean the difference of velocity caused by the action of the force F on a mass free to move, whether accelerating or retarding.

§ 10. There is nothing in my treatise on Dynamics which contradicts your mathematical display. You will find these formulas in my "Elements of Mechanics."

§ 11. Your professorship is not invested with a prerogative to admit or dismiss the force of a moving body; for however arbitrary the force and time may be, they are there, in defiance of your opinion.

§ 12. In the argument referred to there was no call for the term you say I omitted; you will find that term in my "Elements of Mechanics."

§ 13. I hope you will not attempt to introduce any more confusion in Dynamics, such as the term "work of motion," which indicates that motion is a function consisting of work and something else. You have not defined the constituent elements of motion.

§ 14. I do not designate $\frac{1}{2} M V^2$ as "quantity of motion," but have rejected that term in dynamics. Nor should the term "quantity of motion" designate the momentum $M V$. I use only one definite term for each quantity in Dynamics, but you do not appear to have a definite dynamical language.

§ 15. The term "acting force" conveys the idea that there may exist forces which do not act. The simple term "force" implies that it acts, for which reason I proposed to reject "acting." "Motive force" is the proper term for your illustration, but we may call F the acting force and $F - F'$ the motive force. This motive force may be wholly applied against the friction of the car moving with a uniform velocity on the road, or a part of it may be expended in accelerating the velocity of the car. It is not wrong to add the verb "acting" to the term force, but I only proposed to reject the term as superfluous in the sense in which it is often used.

All your forces $F \dot{F}'$ and f are "acting forces" as well as simple "forces." You have not given any example of forces which do not act. It is necessary in Mechanics to distinguish "motive force" from "static force," but both of them are acting.

The purpose for which a force is applied does not alter the nature of that force. Deforming force!!!

(deforming or pressing) force. That it must not be confounded with a pull or a pressure.

§ 16. I consider T, S, F, M as elements.

$$\left.\begin{array}{lll} \mathfrak{V}=\dfrac{S}{T} & \text{in general} & \mathfrak{V}=\dfrac{\partial s}{\partial t} \\ K=FS & \text{"} \quad \text{"} & K=\int F\partial s. \\ \mathfrak{P}=\dfrac{K}{T} & \text{"} \quad \text{"} & \mathfrak{P}=\int\dfrac{F\,\partial s}{T}. \end{array}\right\} \text{Functions.}$$

Also, the mean force $Fm=\int\dfrac{F\,\partial s}{S}.$

Power, $\mathfrak{P}=\dfrac{Fm\,S}{T}=Fm\,Vm.$

§ 17. It is certainly more natural to consider s and t as elements and the differential quotient $V=\dfrac{\partial s}{\partial t}$ as a derived equation than regard t and $\mathfrak{V}$ as elements and $S=\int\mathfrak{V}\,\partial t$ as a derived function.

§ 18. The following are other functions.

The acceleration of motion by the accelerating force,

$$\therefore g'=\frac{F}{M}=\frac{\partial v}{\partial t}=\frac{\partial^2 s}{\partial t^2}.$$

§ 19. The "quantity of motion" $=MV$, and the stored-up "working force" (living force) $\frac{1}{2}MV^2=W\dfrac{V^2}{2g}.$

§ 20. You do not think it right that all authorities without exception should consider "work" $K=\int F\;\partial s$ as independent of time. You will, however, most surely admit that in a finished building there is contained a fixed quantity of work, to do which, of course, some, but an indeterminate, time would be necessary.

§ 21. Consequently we cannot say that the determinable work is dependent on the indeterminable time.

§ 22. If the work was built in a year, it has been done "intensely" (intensive). If three years have been needed for the same work, then it has been done with "less intensity."

The definition of a physical element is, *an essential principle which cannot be resolved into two or more different principles.* Therefore an element cannot be divided by an element and the quotient become a function, as appears in your notions of elements and functions. You say time and space are elements, and then divide space by time and say the quotient is a function—velocity.

When velocity $V=\frac{S}{T}$, we have space $S=V\ T$, which proves that space is a function of velocity and time.

§ 17. Physical facts are not always natural to the mind. There was a time when matter was supposed to consist of only three simple elements—namely, air, water and earth—which was natural in those days.

§ 18. No, sir. These quantities are neither elements nor functions, for they only express the numerical ratio of force and mass.

§ 19. This has been commented on before. Working force means motive force. There is no living force in a dead body.

§ 20. Most decidedly, because the time is included in the space $S=V\ T$. I admit that a fixed quantity of work is required for erecting a building; but when you add the time necessary for it, it cannot be independent of time. If the building can be erected in no time, then that work is independent of time.

§ 21. Work does not bear any fixed relation between its elements, but the product $F\ V\ T$ is work. You say, § 2, that $F\ V=\frac{K}{T}$, from which we have the work $K=F\ V\ T$.

§ 22. Here you introduce a new term, which you have not defined. Is "intensity" an element or a function? If a function, of what elements is "intensity" composed?

§ 23. In this case your formula is right, but your argument is wrong. You eliminate the time from the work in order to get the power. By the term "intensity" you mean power, and from your own formula—

§ 24. We have the work $K=\mathfrak{P}\ T$, which means that the work can be accomplished in any desired length of time, but only at the expense of power.

§ 25. Such is the case with the locksmiths—namely, that one worked with double the power of the other, and consequently earned double the wages in equal lengths of time.

§ 26. Money is equivalent to work, and you must expend $F\ V\ T$ to earn it. There is no fixed relation between F, V and T, but can

§ 23. Not the work but the "intensity of the work," the "arbeitstärke" (working-strength) $\mathfrak{P} = \frac{K}{T}$ depends on the time.

§ 24. If two locksmiths do the same work, the one, however, in half the time the other takes, then the first one has worked with twice the intensity the other did.

§ 25. They received the same compensation for the same work, but the skillful workman received double the wages in the same time because his "arbeitstärke" (working-strength) was double as great.

§ 26. The pay per piece in like work is independent of time, but the resulting earnings per day are in direct ratio to the arbeitstärke (working-strength).

§ 27. The following function may be derived from the pay per piece L and from the time used per piece:

Pay in a unit of time $A = \frac{L}{T}$.

§ 28. According to your idea, on the contrary, the price per piece L would be a function only because it is the product of A and T, and because you will only consider a product, and not a quotient, as a derived function.

§ 29. Such a confusion of ideas as is the case in all the articles concerning "force of falling bodies," especially on page 19 of the *Scientific American* of the 22d of June, 1872, occurs seldom in Germany.

§ 30. There does not exist any "force of falling bodies," only a "bervegungs-arbeit" (work of motion) $= \frac{1}{2} MV^2 = W \frac{V^2}{2g}$, stored up in the falling body, equal to the "verschiebungs-arbeit" (pushing or pulling work) WS, which was necessary to raise the weight W to a height $S = \frac{V^2}{2g}$.

§ 31. This stored-up "external work of motion" is then changed into "verschiebungs-arbeit" (pushing work) $= Rs$ as a mean resistance, R has been overcome through the distance s. Therefore you state correctly that $Rs = WS$. But R is not the force of the falling body, but rather the resistance of the down-pressing body through the distance.

§ 32. Your equation 14 $K = F\,V\,T = \frac{Mm}{V\,T}$, on page 21 of this treatise, is incorrect, as V is the *mean* velocity and F the *initial* force.

vary *ad libitum*, only that their product must correspond with the money.

What you call "strength of work," intensity, or "working strength" is power $\mathfrak{P} = F\ V$.

§ 27. The pay A per unit of time, according to the power of the workman, may be expressed as follows:

$$\text{Wages,} \qquad A = \frac{K}{T} = \mathfrak{P}.$$

§ 28. I have distinguished the terms "element" and "function" by proper definitions, but you use those terms promiscuously according to individual caprice. I maintain that the product of two or more elements is a function, and that a quotient is a solution of a function.

§ 29. The confusions you allude to are written by Dr. Van der Weyde and other doctors of philosophy, for which I am not responsible. I do not consider your ideas of Dynamics to be much better than those of the other professors who have commented upon that subject.

§ 30. Place yourself under a falling body and let it strike upon your head; and if you experience no force, then there is no force in a falling body. Please let me hear from you after you have made the experiment.

§ 31. Is the external work of motion stored upon the surface of the body? The pushing work must then be the internal work, which leaks out when the body strikes?

No force can be experienced without an equal amount of resistance, and the force of a falling body is equal to the force of resistance it meets with.

§ 32. Here you have really discovered an error of mine, for which I am glad to give you due credit, and thank you for calling my attention to it. My idea was to express the work of attraction of two bodies very far apart in space compared with the distance between their centres of gravity when in contact, in which case the force of attraction varies inversely as the square of the distance between the approaching bodies. Your formulas do not include the requisite elements for that work, but merely give the work of a falling body near the surface of the earth.

M and m = masses of the respective bodies.

D = distance apart in feet from which the work is counted.

d = any shorter distance until in contact.

$\varphi = 28693080$, coefficient of attraction.

If $W = m, g$ is the weight of a body at the surface of the earth of a radius a, then the attraction of gravity for the distance x is

$$F = W\frac{a^2}{x^2} = m\,g\frac{a^2}{x^2},$$

and the work K for a fall from the height $x > a$ to the surface of the earth will be $K = -\int F\,\partial x = -m\,g\,a^2\int_x^a \frac{\partial x}{x^2}.$

$$= m\,g\,a^2\left(\frac{1}{a} - \frac{1}{x}\right) = W\,a\left(1 - \frac{a}{x}\right).$$

If x is only larger than a by a very small quantity h, then will

$$\frac{a}{x} = \frac{a}{a+x} = \frac{1}{1+\frac{h}{a}} = 1 - \frac{h}{a} \quad \text{or} \quad 1 - \frac{a}{x} = \frac{h}{a}.$$

Therefore, $K = W\,a\frac{h}{a} = W\,h$, our well-known equation.

§ 33. All German professors are most probably of the opinion that the professor's opinion (page 4) in the main is perfectly correct, and that your answer is composed of sophisms.

§ 34. Willingly, however, do I acknowledge as commendable your desire to arrive at a determination of the dynamical terms, and to eradicate all superfluous ones.

§ 35. The expression, "principle of conservation of force" (princip der erhaltung der kraft), is a very unfortunate one, and unhappily has already led many half-educated persons astray. That chosen by Professor Mach, of Prague, is more correct—namely, "principle of the conversation of work" (princip der erhaltung der arbeit)—and still more correct would be "principle of conversion of work."

§ 36. I therefore say there are four kinds of work which are introconvertible.

First. External pushing or pulling work (aussere verschiebungs arbeit).

Second. External work of motion (aussere bervegungs arbeit).

$$\text{type}: \tfrac{1}{2}\,M\,V^2 = W\frac{V^2}{2\,g}.$$

K = work of attraction in foot-pounds, in drawing the bodies together.

$$\partial K = \frac{M m \partial d}{\varphi d^2} \qquad K = \frac{M m}{\varphi}\left(\frac{1}{d} - \frac{1}{D}\right).$$

This formula expresses the true work in foot-pounds, English measures.

In the case of meteors falling on the surface of the earth we may assume

$$D = \infty \text{ and } \frac{1}{D} = 0.$$

d = 20,887,680 feet radius of the earth.

M = 402,735,000,000,000,000,000,000 matts, mass of the earth.

m = mass of the falling meteor expressed in matts.

The work in foot-pounds of a meteor striking the earth will then be

$$K = 671926000\ m.$$

For very small meteors the greatest part of this work may be converted into heat in passing through the atmosphere, and we call it shooting-stars.

Assuming the mean height of the atmosphere to be 60158 feet, the radius of the atmospheric sphere is 20947018 feet = d.

The velocity with which a meteor enters the atmosphere will then be

$$V = \sqrt{\frac{2\ M}{\varphi\ d}} = 36607.46 \text{ feet per second.}$$

§ 33. I consider it doubtful that all, or even a majority, and not one of the German professors who understood the subject, would be of the opinion of the professor in question. You will no doubt say that my answers to you are composed of sophisms, but I can stand that easily, being accustomed to such charges.

§ 34. I am very glad that you consider my labor commendable, and would state my acknowledgment in emphatic terms but for your employment of such a conglomeration of dynamical terms, which are the worst I have met with.

§ 35. These terms are all useless, and should never be admitted into any school or any text-book. Work in dynamics corresponds to volume in geometry, but we do not give different names to that volume according to the shape of the space it occupies. A vessel holding 100 gallons of water is a fixed volume independent of the shape of the vessel. If the vessel is cylindrical, we do not say it con-

Third. Internal pushing or pulling work at work of pressure (Inner *verschiebungs arbeit* oder *deformerings arbeit*).

As, for instance, in the bent bow, or in an extended or compressed spring, in consequence of the change in the relative position of the molecules, which is against the molecular forces. In permanent gases this is infinitely small, and in condensible vapors it is also very small.

Fourth. Internal work of motion (Inner berveguns arbeit), which appears as heat.

Internal (modicular) work of motion is stored up in a compressed gas or vapor, which can partly change itself into external pushing or pulling work.

§ 37. There is likewise internal work of motion stored in hot gases, the products of combustion, which is transmitted to the water by the heating surface of the steam-boiler, and then changes itself into the internal pushing or pulling force, which must be furnished for the tearing asunder of the molecules of water, and changes also into internal work of motion, which the now generated molecules of steam possess.

§ 38. In forging, rolling, drilling, planing, etc., the greatest part of the work is changed into internal work of motion (heat).

§ 40. Hoping that you will not take my frank remarks on your work in an unfriendly manner, I subscribe myself

Yours respectfully,

Gustav Schmidt,

Professor of Technical Mechanics and of Theoretical Mechanical Engineering at the K. K. German Polytechnic Institute of the Kingdom of Bohemia, Austria.

Prague, July 1, 1875.

The translation of Professor Schmidt's papers was made by Mr. P. Pistor of Philadelphia.

From the foregoing discussion it is clear that the subject of Dynamics lacks perspicuity in the German language for the want of a definite term for the function power.

The term *force* ought to be introduced into the German and Scandinavian languages, leaving the term *kraft* to denote power.

tains 100 cylindrical gallons. So it should be with designation of work, not to give different names to the work according to the proportion of its constituent elements.

It is customary to distinguish indoor work from outdoor work, but in Dynamics it is all $F\ V\ T$.

§ 36. There exists only one kind of work in Dynamics—namely, the product of the three simple physical elements, *force*, *velocity* and *time*.

I should like you very much to go to a machine-shop and explain practically to the workmen, foremen and superintendent your nomenclature of work; and if you can make them understand and appreciate it without laughing at you, I am very much mistaken.

Heat is convertible into work, and consequently must consist of $F\ V\ T$, which is actually the case. The force F is represented by the temperature of the heat and $V\ T$ by the space it occupies in the gas or vapor.

§ 37. The act of combustion is power, which multiplied by time is work; also, the act of evaporation is power, which multiplied by time is work; but in both cases the work of the heat is simply $K = F\ V\ T$.

It is immaterial whether you call it external, internal or infernal work, it is still $K = F\ V\ T$, and nothing else.

§ 38. Your classification of work is not accompanied with the requisite definitions to render your argument admissible.

§ 40. I beg you to accept my sincere thanks for your frank and unsparing remarks on my work. You have liberally furnished precisely what I wanted and asked for in order to test the validity of my reorganization of Dynamics.

In discussions of this kind it is necessary to be frank and free the mind from fiction, for otherwise we could not rightly understand one another, and the interest of science, which we both have at heart, would suffer, notwithstanding our different and even discordant views.

In conclusion let me hope that none of my expressions be interpreted into a want of kind and courteous feeling toward your personality, and I remain, with great consideration,

Yours respectfully,

JOHN W. NYSTROM,

Civil Engineer.

1010 Spruce Street,

Philadelphia, Sept. 1, 1875.

In the English translation of Weisbach's Mechanics, the term and function "power," which is one of the most important functions in Dynamics, does not appear. Even the term "horse-power" is omitted, and cannot be found in the index of that book which otherwise abounds in terms and expressions like those of Professor Schmidt.

On pages 15 and 16 are given a number of rejected terms, which are considered superfluous and confusing in the language of mechanics.

This kind of terms are limited only to books and schools, where they burden the student and tax his time and mind to no purpose, but only to be forgotten when he finds no equivalent for them in practice.

The crowd of subjects which engross the brief years of a school career exact a severe economy of time and labor by the student. It becomes a paramount consideration, therefore, that his acquirements should in his subsequent experience be found to possess an unequivocal practical value, which has heretofore not been fully realized.

A graduated student of Mechanics, although expected to be well versed in that subject, is, when brought to a practical test, often found wanting, as is shown in periodicals of the day, where we rarely find a sound article on Dynamics. For example, in the London *Engineer* lately appeared an article on Dynamics of heavy ordnance, written by an English artillery officer, stating that

$$\text{" The energy in } \textit{vis viva} \text{ in pounds} = \frac{W V^2}{2 g}\text{,"}$$

$$\text{whereas it is not pounds, but work} = \frac{W V^2}{2 g}.$$

This function is called "energy" by doctors of philosophy, who very often represent it as a very mysterious phenomenon.

The term " energy" is not used in the English translation of Weisbach, except in a note by the translator.

The term " energy" is derived from the Greek ἐν-ἔργου, of which ἐν means inner or within, and ἔργου means work.

"Kinetic energy" (κινητος-ἔργου) means moving energy.

"Potential energy" (Latin, *potentalis*) means powerful energy.

These terms and expressions have originated at times when the science of Dynamics was in a very clouded condition, and have since been retained with various kinds of conflicting definitions.

The sense in which the term "energy" is generally used, means simply "work," which consists of only $F\ V\ T$, and nothing more or less.

In the formula $\frac{W V^2}{2\,g}$, V means the final velocity of a falling body, which is double the mean velocity of the fall. W = force of gravity F, and $T = \frac{V}{g}$, the time in seconds of the fall, of which $V = g\ T$.

Energy or work, $$K = \frac{W\,V^2}{2\,g} = \frac{W\,V\,T}{2} = \tfrac{1}{2}\,F\,V\,T.$$

It is simply the force F of gravity which accomplished the work K of the falling body, giving it a velocity V in the time T.

There exists no such distinction as inner or outer energy or work, nor kinetic or potential energy, which are all simply work $K = F\ V\ T$.

When a reader attempts to gather information from a book with those high-sounding terms, he may be impressed with the idea that the subject is much too profound for him to learn, and that he has not sufficient intellect to grasp it, whilst the fact is that there is nothing in it but simply $F\ V\ T$.

One evil of high-sounding terms is that they are often sophistically and successfully used for delusion, of which the writer could refer to many cases, but fears that in so doing his motives would be misunderstood.

On one occasion a professor whilst arguing the subject of radiation of heat spoke about "dynamical temperature, statical temperature, potential temperature and actual temperature." On being asked "What is the difference between potential and actual temperatures?" the professor answered, "Potential temperature refers to volume."

Question. "Is potential temperature measured by a thermometer?"

The professor could not answer, but gave it up.

High-sounding terms, in fact, serve the same purpose as feathers of many colors in a hat—namely, to decorate the subject.

www.ingramcontent.com/pod-product-compliance
Lightning Source LLC
LaVergne TN
LVHW011224110826
845150LV00006B/1542

9781425516154